W9-BER-178
FREEPORT
MEMORIAL LIBRARY
Presented by
LANAI HOUSE
1995

ISLANDS

ISLANDS

The Illustrated Library of the Earth

CONSULTING EDITORS

ROBERT E. STEVENSON, Ph.D.

FRANK H. TALBOT, Ph.D.

RODALE PRESS
EMMAUS, PENNSYLVANIA

Published 1994 by Rodale Press, Inc.
33 East Minor Street, Emmaus, PA 18098, USA

By arrangement with Weldon Owen
Conceived and produced by Weldon Owen Pty Limited
43 Victoria Street, McMahons Point, NSW, 2060, Australia
Fax (02) 929 8352
A member of the Weldon International Group of Companies
Sydney ● San Francisco ● London

CHAIRMAN: Kevin Weldon
PRESIDENT: John Owen
GENERAL MANAGER: Stuart Laurence
PUBLISHER: Sheena Coupe
SERIES COORDINATOR: Tracy Tucker
ASSISTANT EDITOR: Julia Burke
COPY EDITOR: Mary Rennie
SCIENTIFIC ADVISOR: Terence Lindsey
DESIGNER: Toni Hope-Caten
DESIGN CONCEPT: Andi Cole, Andi Cole Design
COMPUTER LAYOUT: Gary Fletcher
PICTURE RESEARCH: Julia Burke, Joanna Collard
ILLUSTRATION RESEARCH: Joanna Collard
CAPTIONS: Terence Lindsey
INDEX: Diane Harriman
ILLUSTRATIONS AND MAPS: Anne Bowman, Jon Gittoes, Mike Gorman, David Kirshner, Nicola Oram, Oliver Rennert, Patrick Watson
COEDITIONS DIRECTOR: Derek Barton
PRODUCTION DIRECTOR: Mick Bagnato
PRODUCTION COORDINATOR: Simone Perryman

Library of Congress Cataloging-in-Publication Data

Islands/consulting editors, Robert E. Stevenson, Frank H. Talbot.
p. cm.—(The Illustrated library of the earth)
Includes index.
ISBN 0-87596-632-2: hardcover
1. Islands. I. Talbot, Frank. II. Series
GB471.I85 1994
508.3142—dc20 93-36352
CIP

If you have any questions or comments concerning this book, please write to:
Rodale Press
Book Readers' Service
33 East Minor Street
Emmaus, PA 18098

Manufactured by Mandarin Offset
Printed in Hong Kong

Distributed in the book trade by St. Martin's Press

10 9 8 7 6 5 4 3 2 1

A WELDON OWEN PRODUCTION

JACKET: *Kuta Bandos beach, Maldives. Photo by* ZEFA*/Australian Picture Library.* JACKET INSET: *A silhouette of a snowy egret at dawn, Florida, USA. Photo by Lynn M. Stone/The Image Bank.* ENDPAPERS: *Gastropods and bivalves washed ashore by a change in tide. Photo by Jean-Paul Ferrero/*AUSCAPE. PAGE 1: *A Huli tribesman, Papua New Guinea. Photo by Kevin Deacon/Dive 2000.* PAGE 2: *Sunset on a tropical island, Praslin, Seychelles. Photo by Hans-Peter Merten/Bruce Coleman Ltd.* PAGE 3: *The Fijian archipelago consists of more than 800 islands and islets, including the coral islet of Malamata. Photo by Jean-Paul Ferrero/*AUSCAPE. PAGES 4–5: *An island endemic: the chameleon* Chameleon lateralis *of Madagascar.* PAGES 6–7: *Two wandering albatrosses* Diomedea exulans *in courtship display, South Georgia.* PAGE 8: *A brown lemur* Lemur fulvus *stoops to drink, Madagascar. Photo by Frans Lanting/Minden Pictures.* PAGES 10–11: *A New Guinea native in full ceremonial paint and makeup.* PAGES 12–13: *Lord Howe Island in the Tasman Sea is fringed by a thriving coral reef fish community.* PAGES 52–53: *A marine iguana* Amblyrhynchus cristatus *with a Sally Lightfoot crab, Hood Island, Galapagos.* PAGES 108–109: *Chimbu highlanders, Papua New Guinea.*

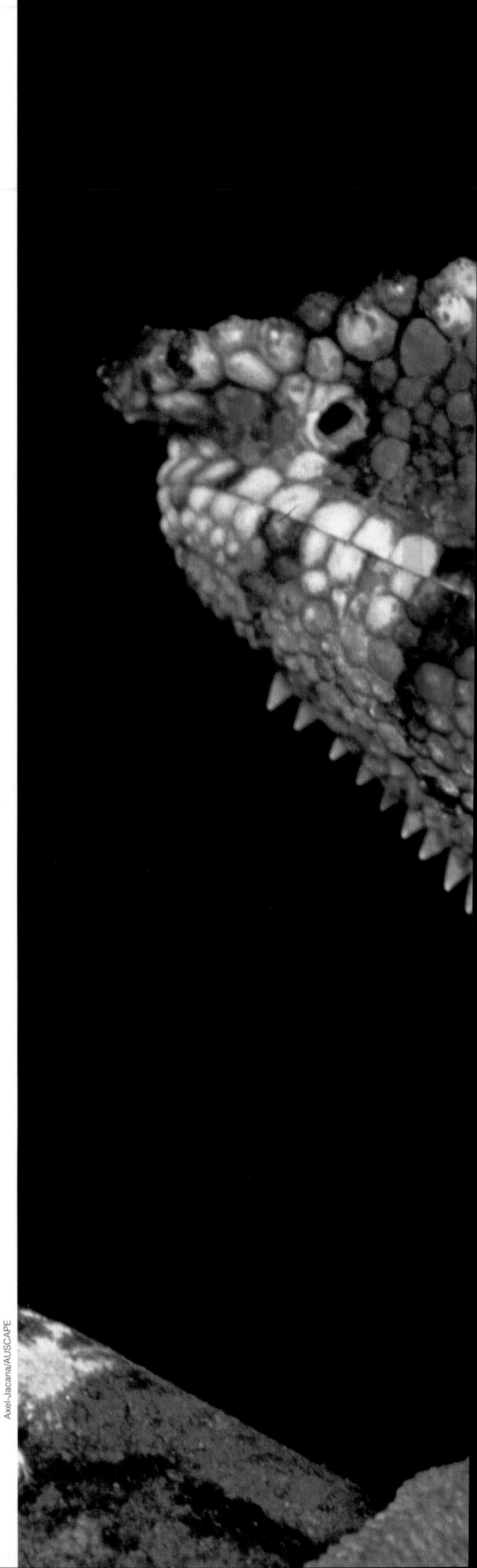
Axel-Jacana/AUSCAPE

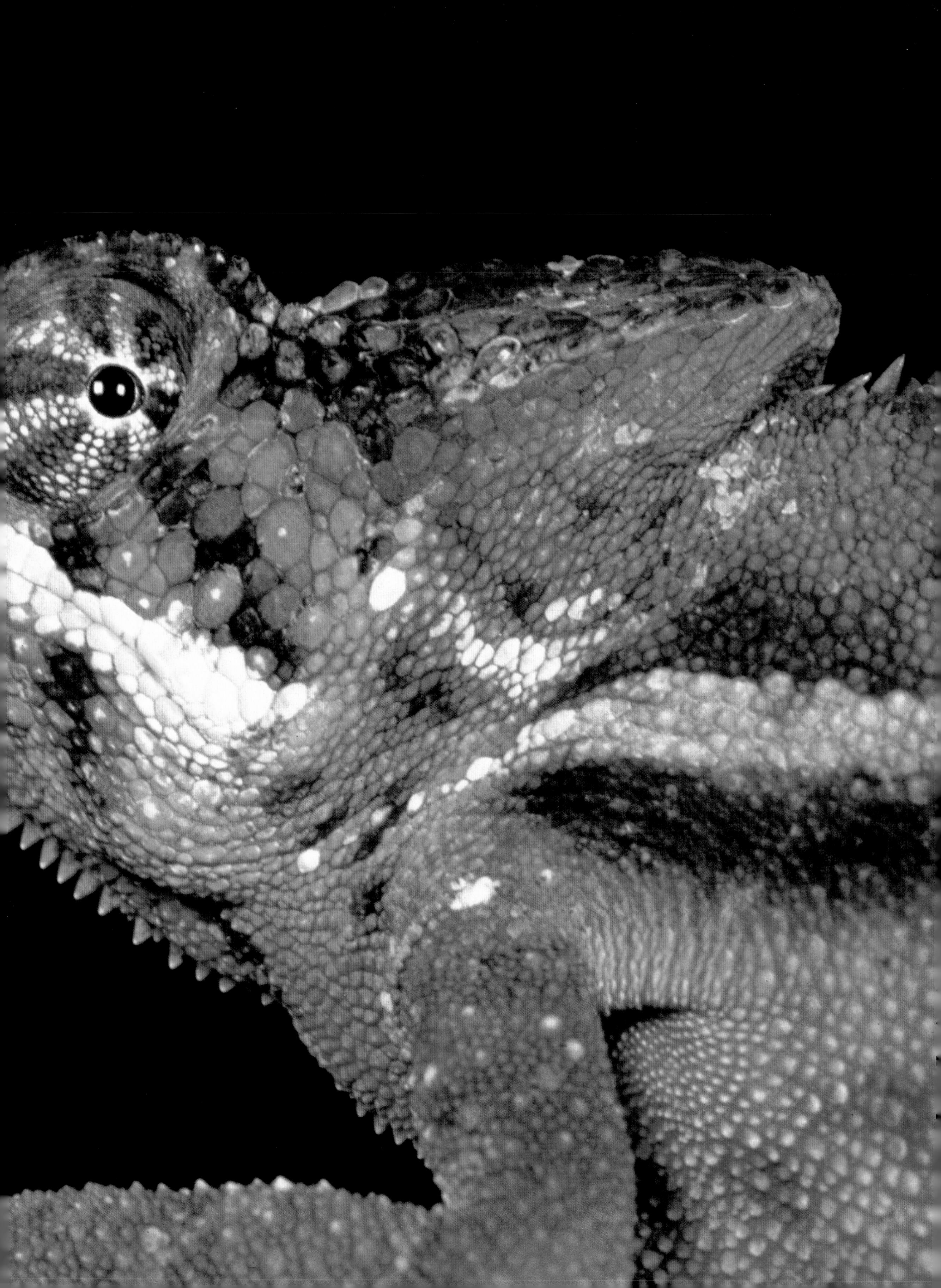

Contributors

DR. RICHARD S. FISKE
National Museum of Natural History, Smithsonian Institution, Washington DC, USA

ASSOCIATE PROFESSOR PETER G. FLOOD
Department of Geology and Geophysics, University of New England, Armidale, Australia

ASSOCIATE PROFESSOR BENT FREDSKILD
Botanical Museum, University of Copenhagen, Denmark

DR. STEPHEN GARNETT
Queensland Department of Environment and Heritage, Australia

DR. RICHARD W. GRIGG
Department of Oceanography, University of Hawaii at Manoa, Honolulu, USA

PROFESSOR HAROLD HEATWOLE
Department of Zoology, North Carolina State University, Raleigh, USA

DR. DAVID HOPLEY
Sir George Fisher Centre for Tropical Marine Studies, James Cook University of North Queensland, Townsville, Australia

STUART INDER MBE
formerly Publisher of *Pacific Islands Monthly* and *Pacific Islands Year Book*, Sydney, Australia

PROFESSOR E. ALISON KAY
Department of Zoology, University of Hawaii at Manoa, Honolulu, USA

DR. KNOWLES KERRY
Australian Antarctic Division, Australia

TERENCE LINDSEY
Associate of the Australian Museum, Sydney, Australia

ASSOCIATE PROFESSOR KENNETH McPHERSON
Indian Ocean Centre for Peace Studies, Curtin University and University of Western Australia, Perth, Australia

PROFESSOR ALEXANDER MALAHOFF
Department of Oceanography, University of Hawaii at Manoa, Honolulu, USA

PROFESSOR SIDNEY W. MINTZ
Department of Anthropology, Johns Hopkins University, Baltimore, USA

DR. STORRS L. OLSON
National Museum of Natural History, Smithsonian Institution, Washington DC, USA

DEBORAH ROWLEY-CONWY
Arts, Libraries and Museums Section, Durham County Council, UK

DR. PETER ROWLEY-CONWY
Department of Archaeology, University of Durham, UK

DR. FRANK H. TALBOT
National Museum of Natural History, Smithsonian Institution, Washington DC, USA

DR. PAUL MICHAEL TAYLOR
National Museum of Natural History, Smithsonian Institution, Washington DC, USA

DR. DIANA WALKER
Department of Marine Botany, University of Western Australia, Perth, Australia

DR. G.M. WELLINGTON
Department of Biology, University of Houston, USA

DR. JOHN C. YALDWYN
National Museum of New Zealand, Wellington, New Zealand

Jean-Paul Ferrero/AUSCAPE

CONTENTS

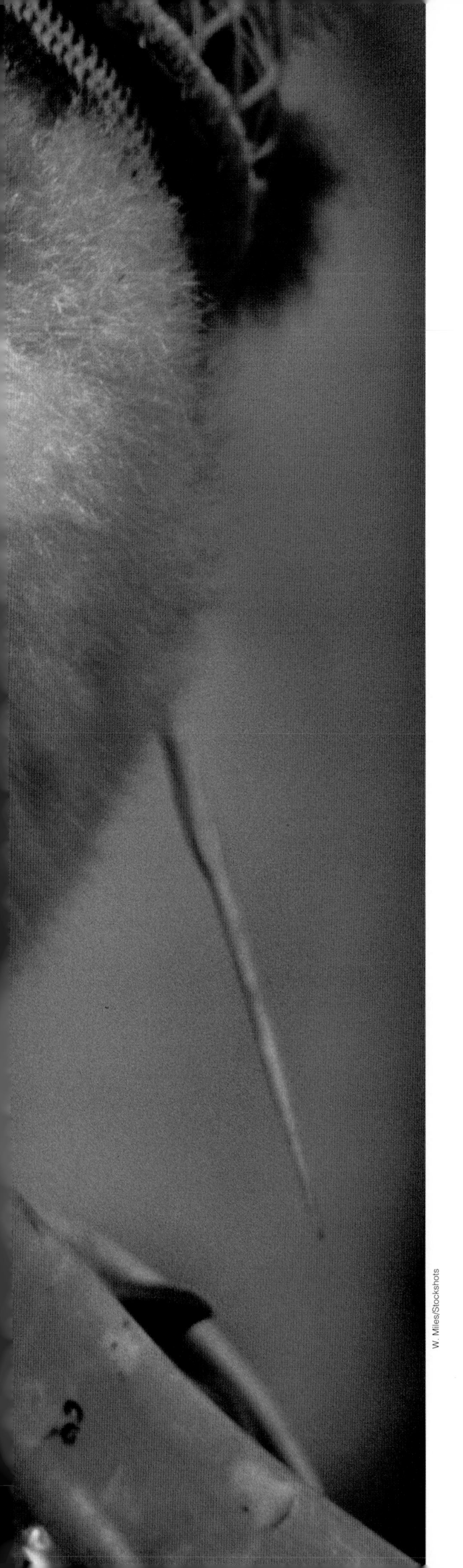
W. Miles/Stockshots

Foreword

Islands are everywhere: in ponds, lakes, rivers, inland seas, and the great expanses of the oceans. Some islands are the centers of human empires; many others have made a graceful home for humankind as well as the indigenous flora and fauna that evolved on their grounds; still others remain inhospitable and pristine through the ages. Humankind has so totally changed the environment of some islands—such as Manhattan, New York—that they hardly qualify as islands. On others, humans have worked hand-in-hand with nature to find a harmonious balance between the material needs of our species and the environment.

In their infinite variety and forms, islands provide microcosms of all the world's environments. They represent, therefore, laboratories where we can study every natural interaction. The very birth of an island and its colonization by the first organic forms to reach its shores can be studied on new islands that rise out of the sea, such as those off Hawaii and Iceland.

We have seen too the effect of climatic changes on islands. At the end of the last ice age, about 18,000 years ago, the sea began to rise from just more than 100 meters (330 feet) below its present level, and the coastlines of islands and continents alike were gradually changed by erosion and deposition, until the sea reached its present "near stand-still" about 2,500 years ago. We can observe the response of indigenous animals, plants, and human populations to short-lived climatic changes: for example, the effect of hurricanes can be analyzed in hundreds of locations in the Pacific and Indian oceans. On islands off Chile, Canada, and Greenland, we are able to evaluate the influences of ice, snow, and glaciers on a habitat.

Some islands have come to their end in grand tectonic dramas, such as the gigantic explosion of Krakatoa in 1883. Other volcanic islands, Hawaii and Iceland, for example, are growing. Even Scotland, with its magnificent firths, continues to rebound from its great ice-age glacier.

Our discussions herein cover three types of islands: continental, volcanic, and coral islands. In each case, the type of island provides the basic information of the geological forces that created it, the structure and stability of the sea floor on which it lies, and the oceanic and atmospheric forces that have modified it. Islands, uniquely attractive, even mysterious, and separated in some cases by thousands of kilometers, have environments that are closed systems and, therefore, vulnerable. As a result, they present to us the most exciting research laboratories imaginable. To maintain their pleasures, they must be sustainably managed.

ROBERT E. STEVENSON
Secretary-General, International Association for
the Physical Sciences of the Ocean, California, USA

PART ONE

The Island Realm

Behind the apparent scattering of islands around the world lies a unifying geological story. Some islands have their origins in the workings of titanic forces in the Earth's crust, others in the restless movements of the ocean.

1 WORLDS APART

14

Islands are dispersed throughout the world's oceans. They may lie close to the continent of which they were once part or in the open ocean far from any substantial landmass. They may range in size from huge landmasses, such as Greenland, to rocky outcrops no more than a few square meters in area.

2 CONTINENTAL ISLANDS

24

Continental islands may be fragments of continental plates or drowned coastal ranges isolated by changes in sea level. Others form from sediments brought to the sea by rivers and glaciers, swept into shifting heaps by ocean winds and currents.

3 ISLANDS OF FIRE

34

Born of fiery fountains of molten magma welling from deep within the Earth, volcanoes may grow to the surface of the sea and beyond to form islands. In time they may collapse under their own weight, and eventually subside beneath the surface.

4 THE FORMATION OF CORAL ISLANDS

46

In the warm waters that surround the equator, islands may be built up, almost grain by grain, from the skeletons of coral polyps, each new generation growing upon the skeletal remains of the previous one.

1 Worlds Apart

DAVID HOPLEY

An island is any landmass smaller than a continent, which is completely surrounded by water. This water can be the open ocean, the shallow seas of continental shelves, the mouths of estuaries, or even freshwater rivers and lakes. Indeed, the term "island" applies equally to ephemeral dune islands, the mountains of rock erupted from fiery volcanoes, and the fragments of ancient continents.

M. le Coz/EXPLORER-AUSCAPE

Islands range in size from Greenland's 2.1 million square kilometers (810,800 square miles) to minute rocky fragments, like that serving as foundation for this exposed lighthouse off the coast of Brittany, France.

Rocks and Large Places

Australia, at 7.6 million square kilometers (almost 3 million square miles) is sometimes referred to as the island continent and is the largest isolated landmass in the world. Its unique plants and animals are the result of evolution during almost 100 million years of separation from any other continent.

Beyond this exceptional example, the largest island is Greenland in the north Atlantic, which has an area of 2.1 million square kilometers (810,800 square miles). Other prominent islands include New Guinea at 820,660 square kilometers (316,860 square miles), Borneo at 746,550 square kilometers (288,240 square miles), Madagascar at 587,040 square kilometers (226,660 square miles), and Baffin Island, Canada, at 476,070 square kilometers (183,810 square miles). These are major landmasses, which are put into perspective by comparison with the area of all the British Isles: 314,780 square kilometers (121,540 square miles).

At the other end of the scale, there are isolated rocks and pinnacles no more than a few square meters in area, often washed over by the sea, and only temporarily inhabited by plants and animals. Yet even these islands have importance to humankind: they may cause shipwrecks or their associated shallows may be important fishing grounds. They are also the means of greatly extending the 200 mile national Exclusive Economic Zone (EEZ) that lies off the coast of each country, and over which it has fishing and mineral rights.

The European perception of islands as worlds distant and isolated has changed through time. Locations once regarded as being on the edge of the civilized world are now attractive places for tourists, scientific reserves, or military stations. On Molokai in Hawaii, where lepers were banished in the nineteenth century, multinational hotels now rise and golf courses proliferate. On Norfolk Island, in the Tasman Sea, where prisoners were incarcerated from 1788 to 1855, there is also a flourishing tourist industry.

Even the smallest islands dotted within the vast expanses of ocean have been utilized: as coaling

David Doubilet

stations for refueling ships; as cable stations providing links in worldwide communications; and as airfields in the years when short-haul aircraft were in service for transoceanic flights. Many Pacific atolls provided necessary landing places en route between North America and Asia, New Zealand, and Australia; for more than a century, islands all over the world have been used for navigational beacons; more recently they have served as weather stations and satellite tracking centers.

Island resources have also attracted transient visitors. For early navigators they offered shelter, anchorage, food, and water. For later European mariners, they provided exploitable resources in the form of timber and minerals (including guano fertilizer); animal resources, such as turtle meat, the furs of seals and other aquatic mammals, and oil from muttonbirds. In addition, they provided bases from which to extend into fishing and whale hunting grounds. More recently, many islands have become the

Truly isolated islands are unusual: mostly they are distributed in archipelagos—clusters of small islands like the Duke of York group, Papua New Guinea, seen here from the air.

CONTINENTS IN COLLISION

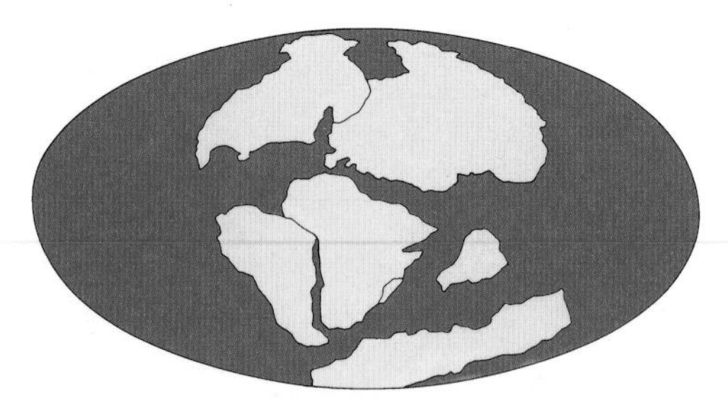

130 million years ago

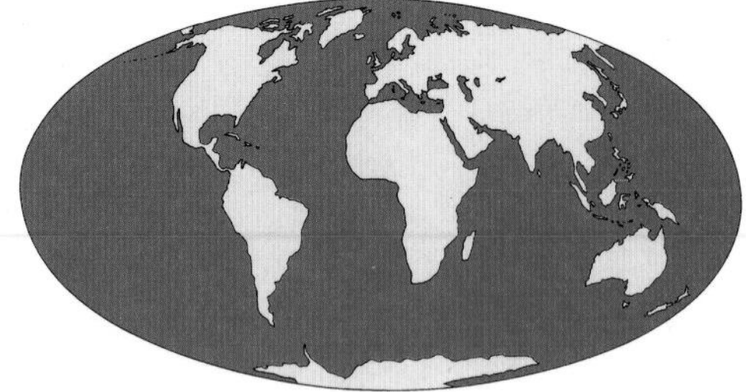

today

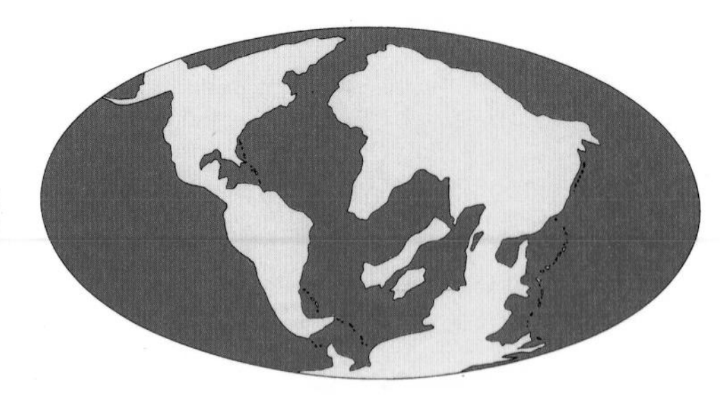

150 million years from now

By 130 million years ago, the great southern landmass of Gondwana had broken away from Pangaea and had itself begun to fragment, the fragments drifting apart to reach the continental distribution we see today. As the plates continue to move, so will the shape of the continents continue to change.

vacation destinations of the traveling public. No other industry has exploited islands to the extent that tourism has over the past 30 years.

First and foremost, however, islands are places where people live. Some, such as the British Isles and Japan, are highly developed nations of many millions of people; at the other end of the scale are the non-industrialized civilizations, such as those of the remote Pacific islands of Polynesia.

While worlds apart, the characteristics of these nations as islands have influenced their history and development. Without metals, and without what we regard as modern technology, the Polynesians were able to navigate most of the vast expanse of the Pacific Ocean. The seafaring qualities of the British were foremost in the development of the United Kingdom as a major world power with colonies extending around the globe.

Sea-floor spreading, modeled in this computer graphic, occurs as molten magma seeps upward from the Earth's interior along mid-oceanic ridges; water cooling forms solid blocks that are slowly forced apart by continued magma upwelling.

Studio R/Japan Broadcasting Corporation

Drifting Lands

In 1912 Dr. Alfred Wegener, a German meteorologist, described how many of the world's continents could be fitted together, like an intricate jigsaw puzzle. He noted, for example, that the east coast of South America could be slotted into the west coast of Africa, and that Australia, India, and the east coast of Africa fitted neatly around Antarctica.

These observations had led him to the idea of continental drift: that the continents, at one stage, had formed a massive supercontinent which came to be called Pangaea—made up of Gondwana in the south and Laurasia in the north—and that subsequently the continents had split apart and drifted to their present positions. The idea originally met with great skepticism, but during the 1950s and 1960s, examination of the oceans by geoscientists gave it new support.

Mid-oceanic areas are characterized by ridges rising from the ocean floor, even though they may never grow to within 1,000 meters (3,300 feet) of the ocean's surface. In contrast, the edges of the oceans are often flanked by very deep trenches that are the deepest parts of the oceans, more than 10,000 meters (33,000 feet) deep. In the early 1960s, research revealed that the ocean floor is young along the oceanic ridges and often accompanied by active submarine volcanic action. The farther away from the ocean ridges it is, the older the ocean floor becomes, until it reaches a maximum age close to the deep trenches.

These observations led to the concept of sea-floor spreading: that along the ocean ridges new crust is constantly being formed, which then moves laterally away from these ridges. The deep trenches, known as subduction zones, are areas where the crust finally plunges back down into the mantle of the Earth. Thus, as new crust is formed, old crust is constantly being consumed.

By delineating the areas of new crustal formation and the long oceanic trenches, it was possible to identify separate crustal plates that form the Earth's dynamic surface. There are six major plates, as well as

Frans Lanting/Minden Pictures

a large number of smaller fragments. Each is moving in a different direction. Some travel in tandem: the San Andreas Fault, for example, is the result of the parallel movement of the Pacific and North American plates. Others come into direct collision with one another. When this happens one plate moves beneath the other to form the deep trenches in subduction zones. An example of this is the northward-moving Indo-Australian Plate, which has collided with the Eurasian Plate along the line of the Himalaya and the Indonesian islands.

Further observations showed that the rocks forming the new crust at the mid-ocean ridges are heavier than rocks that form the major continents. Like pumice stone floating on the sea, the continental rocks of lighter material float as rafts on the denser crust of the oceans. Thus, the continents really are drifting. Ocean spreading occurs at rates of between 3 and 10 centimeters (1 and 4 inches) per year, dating the breakup of Gondwana, the old supercontinent of the Southern Hemisphere, at approximately 130 million years ago. Where continental and oceanic crust collide, invariably the denser oceanic material plunges beneath the lighter continents to form the typical deep trench close to the continent.

Frans Lanting/Minden Pictures

A Madagascan landscape. This large continental fragment supports extremely distinctive plant and animal communities. The eastern half of the island is humid and forested, but arid desert dominates the southwest.

A band of ring-tailed lemurs Lemur catta *basks in Madagascar's evening sun.*

Continental Islands

During his great exploratory voyages in the late 1700s, the navigator Captain James Cook identified two types of island: high islands, rising above the horizon and easily seen from a distance, and low islands associated with coral reefs, and no more than a meter or two above sea level. Today the classification of islands starts with their geological composition and the way they were formed. All islands fall into two main categories: continental and oceanic.

J.M. la Roque/AUSCAPE

Hjalmar R. Bardarson/Oxford Scientific Films

Top. *Not all coral islands are formed as atolls. Some, like the island of Pohnpei, Micronesia, have their origins in the gradual subsidence of barrier reefs.*

Bottom. *Islands are still being formed: the volcanic island of Surtsey off Iceland, for example, emerged during a volcanic eruption in November 1963 and is still growing.*

Opposite. *The youthful character of volcanic islands is often revealed by spectacular cliffs and similar formations, such as the Pinnacles in the Kalalau Valley, Kauai, Hawaii. Here, the forces of wind and water erosion have had relatively little time to exert their influence.*

As the lighter continental crust is rafted around the Earth's surface on the denser oceanic crust, many fragments become isolated as large continental islands. The continent of Australia is the largest of these fragments, but others include islands such as Madagascar, Greenland, New Guinea, and New Zealand.

The rocks of continental islands are made of granites, acid volcanics, and their sedimentary derivatives. They are lighter and more acidic than those of oceanic, or noncontinental, islands, which are made chiefly of volcanic basalt. They may also be substantially older. The combination of greater age and the possibility of former contact with other continental landmasses produces a more diverse and complex biota (range of life-forms) than is ever seen on oceanic islands.

Although the larger islands may not have been part of a continent for more than 100 million years, the numerous smaller continental islands found adjacent to the major continents have had a much more recent connection. Surrounded by shallow water, usually less than 100 meters (330 feet) deep, they have been affected by the major changes in sea level caused by the glaciations of the Quaternary period (the last 2 million years). Each time the Earth cooled, ice built up to form the major ice sheets that covered North America, northwestern Europe, and other parts of the world.

The water that went into these ice sheets came from the oceans, which correspondingly fell by more than 100 meters (330 feet), thereby exposing the continental shelves. What are now islands stood as coastal hills and ranges surrounded by the exposed shelf. As the ice waxed and waned during the past two million years, the continental shelves were alternately drowned and exposed. Many land areas that are now islands were simply extensions of the adjacent continent. Ireland, for example, became isolated from Britain and Europe only in the last 8,500 years, and New Guinea became separated from Australia as recently as 8,000 years ago.

Oceanic Islands

The volcanic activity in areas of mid-oceanic ridges is sometimes sufficient to produce islands. The islands of Tristan da Cunha, Ascension, St Paul Rocks, and the Azores were formed in this way. Iceland, too, is part of a significant mid-ridge volcanic island mass. Sometimes oceanic islands form because of volcanic activity around subduction zones. As one plate slides under another in these areas of collision, the interaction of molten material, gases, and heat can cause eruptions through the overlying plate and water, often resulting in curved rows of islands known as island arcs.

Apart from these areas, there are other very active mid-oceanic volcanic areas. Sometimes referred to as "hot spots", they appear to maintain a relatively stable position with respect to the plates moving over them. The best-known example is the Hawaiian island chain. At the present time, the most active hot spot is located at the southern end of the "Big Island" of Hawaii, at the site of the Kilauea volcano. However, the Pacific Plate has been passing over this spot toward the northwest for more than 30 million years, and the direction of movement can be clearly seen in the alignment of the Hawaiian Islands from southeast to northwest. The farther away they are from the Kilauea hot spot, the older and more eroded the islands become.

The sea floor, too, subsides the farther it travels from the original hot spot. However, as the volcanic rocks sink, coral reefs occupy their flanks and slowly grow upward. Indeed, volcanic rocks can disappear beneath the sea and be covered completely by coral reefs. Thus are coral islands formed. Right at the end of the Hawaiian chain is Pearl and Hermes Reef, and Midway and Kure islands. All lie on volcanic foundations, which are now more than 150 meters (500 feet) below the ocean surface.

Coral reefs form not only on volcanic rock. In the right latitudes, reefs will form on shallow continental shelves; however, maintaining the right temperature for their growth is critical. Kure Atoll is the northernmost atoll in the world. Here, coral reef growth is at its extreme because of the cooler waters outside the tropics. As Kure will move farther northward into cooler waters in the future, coral growth is likely to cease, and the atoll will slowly subside beneath the ocean. This process has already occurred in some places, such as in the line of seamounts known as "guyots" that form the Emperor Seamount chain. These flat-topped features were previously coral reefs that could not keep up with the subsidence of their foundations.

ISLANDS UPLIFTED

Not all atolls and their volcanic foundations are lost forever by subsidence and consummation in the subduction zones. As they approach plate margins, islands and their associated reefs can also be uplifted by the violent tectonic forces in the areas of plate collision. Thus, reefs that were once completely beneath the sea may be uplifted several hundreds of meters to form high limestone islands with distinctive coral-reef terraces. These can be seen on the Trobriand Islands of Papua New Guinea, Makatea in French Polynesia, and some of the Loyalty Islands in New Caledonia.

The distinctive features that are seen on oceanic islands close to plate margins are also seen on continental islands. For example, New Guinea, which forms the bow wave of the Indo-Australian Plate as it plows into the westernmost part of the Pacific Plate, has experienced violent tectonic uplift. On its northern shores, around the Huon Peninsula, coral-reef terraces rise dramatically to an elevation of more than 600 meters (2,000) feet.

In areas of such vertical uplift, a number of former continental islands are now rejoined to their adjacent mainland. Areas once under 3,000 meters (10,000 feet) of ice at the height of the last glaciation are also undergoing rapid uplift as the immense weight of the ice is removed and the surface rebounds. Islands around the Baltic Sea, for example, are being uplifted and joined to the mainland shores of Sweden and Finland.

There are, of course, considerable volumes of ice still covering the Earth's surface, particularly in Antarctica and Greenland: 2.4 percent of the Earth's water is locked up as ice. Were this to melt in the future, it is estimated that the sea level would rise at least 50 meters (160 feet) with drastic effects on all the world's coastal plains. Although some islands would disappear through such a rise in sea level, many others would be formed as coastal ranges and hills were isolated. Indeed, the island status of many landmasses is ephemeral. ■

Frans Lanting/Minden Pictures

WHERE SEA MEETS ROCK

The zone in which the sea makes contact with the rocky shores of an island is a focus for biologists, ecologists, and geologists alike. This zone extends from low-tide level, at the base of effective wave action, through the intertidal zone to the highest levels that spray can reach. On exposed shorelines, with high tidal ranges, this can be more than 100 meters (330 feet).

Starfish are typical animal inhabitants of the realm where ocean and rocky shore meet.

The characteristics of the rocky shore zone depend, to a large extent, on the elevation of the immediate hinterland and the resistance of the rocks to erosion. Whether low rocky shores or coastal cliffs, the erosional processes are the same. Rocks such as chalk and sandstone may be eroded relatively easily by the complex processes operating at their base, resulting in vertical, or even overhanging, cliffs. At the base of cliffs, there may be a near-horizontal rock platform, exposed only at low tide. There is no doubt that large waves breaking on a cliff-face can cause explosive action in cavities, but more persistent erosion is produced by gravels picked up in the waves. These act as abrasive tools which chisel out an intertidal platform. Chemical decomposition, due to constant washing and drying, can produce pitting and honey-combing of rocks on and above the platform and in the notch that can form at the base of cliffs. In high latitudes with cold climates, frost action may break down rock materials and the action of the waves the resulting debris.

Plants and animals can be more directly responsible for the processes of shore platform and notch formation through bioerosion. Cyanobacteria can bore into a rock-face, and other organisms, such as chitons and sea urchins, bore or rasp their way into the intertidal rock for shelter. Many mollusks and echinoids graze on the rich plant life, but in doing so also rasp some of the rock materials from the platform. Although individual animals may remove only small amounts, high populations can result in erosion rates which can be measured in millimeters per year.

The vertical distribution of these bioeroding organisms and other plants and animals which live within the littoral (coastal) zone varies with climate, exposure, and tidal range around the world. Some general features are recognizable. The sublittoral (offshore) zone is constantly under water. Red algae utilizes the part of the light spectrum that penetrates best into water, so it may dominate deep down. Large brown algae and sargassum may be found here also, and huge kelp (brown algae) may be rooted as deep as 15 to 30 meters (50 to 100 feet), with long trailing fronds at the surface. In the intertidal zone, plants are often shorter and may form turfs. Animals in this wave-torn area are usually ones that can hold on tightly.

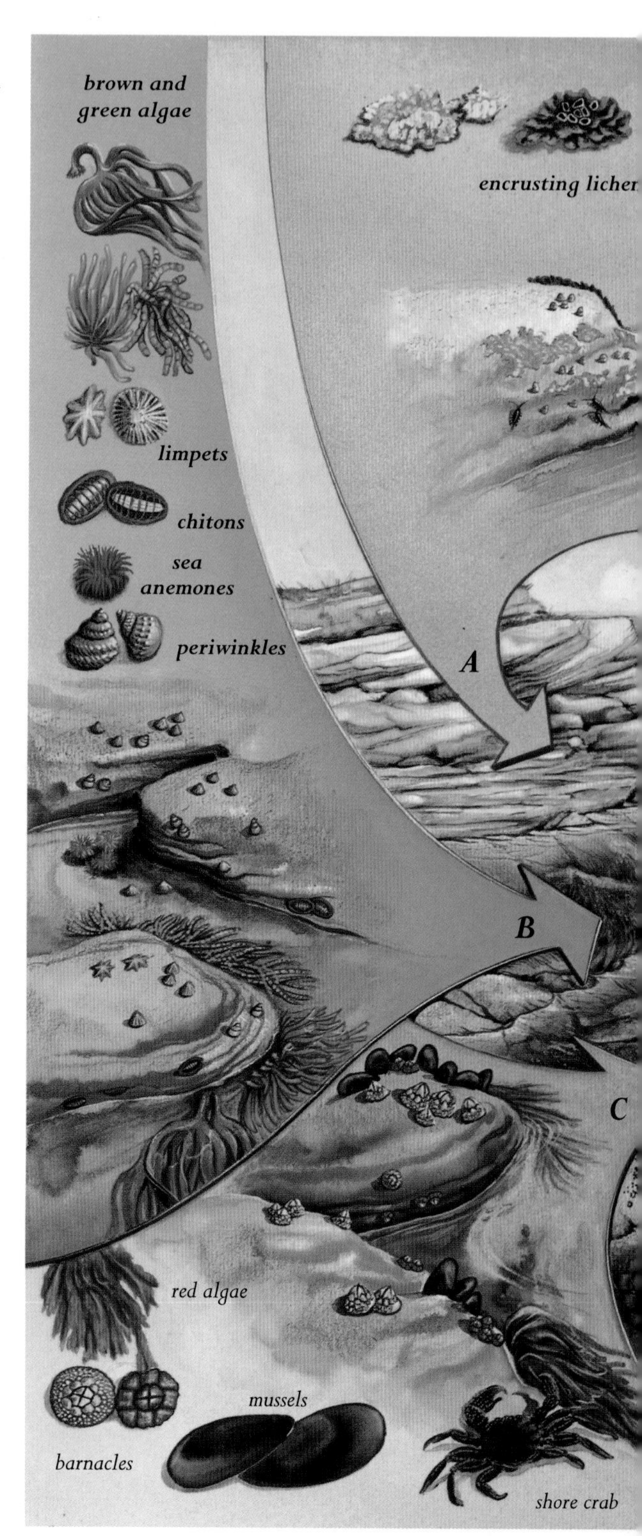

The eulittoral zone is the part of the rocky foreshore which is affected by tidal oscillation. Varying degrees and times of exposure provide distinctive zonation in this area. The uppermost zone, which is uncovered by most outgoing tides, may be dominated by small barnacles; the mid-tide zone by oysters, limpets, and blue-green algae; and the lowest zone, exposed only by the lowest tides, is dominated by aloal mats accompanied by large barnacles. The superlittoral zone, wetted only by extremely high tides or by spray, is dominated by organisms with marine affinities—littorinid snails, patches of cyanobacteria, and at higher levels crustose gray, orange, or yellow lichens.

However, the actual organisms vary worldwide, and in some areas a single species of plant or animal may dominate a zone completely as, for example, with oyster zones. In tropical areas at about the level of spring low tides, corals may become the dominant organism, and the rocky shore may terminate abruptly in a reef-flat. ●

A rich community of plants and animals characterizes the meeting place between the sea and rocky shores. Tidal action results in a spectrum of microhabitats ranging from wet to dry, but four distinct zones can usually be distinguished. This illustration depicts a typical temperate-area shoreline.

A SPLASH ZONE
Occasionally receives spray during stormy weather.

B HIGH-TIDE ZONE
Covered only by the highest tides. During neap tides it may remain dry, apart from spray, for several days.

C INTERTIDAL ZONE
Covered by water on every high tide and above sea level on every low tide.

D LOW-TIDE ZONE
Exposed only on the lowest Spring tides. May remain covered by water for several days at a time.

ISLAND BEACHES

Beaches occur wherever there are small indentations around island shores. Only the steepest and most rocky islands completely lack beaches, and some islands consist entirely of beach material and associated dunes.

Jane Burton/Bruce Coleman Ltd.

A typical shingle beach in southern England.

Center. *Black sand beaches, such as this one in Hawaii, are formed from basalt and often occur on volcanic islands.*

Boulder beaches characterize many high latitude shorelines, where vigorous frost action tends to fracture rocks rather than wear them away.

Beaches are formed by the action of waves as they are refracted around the outline of the island, sweeping sediments into the lowest energy sectors. This sweeping action of the waves makes beaches important pathways for the entry of vegetation into islands, largely as float seeds.

BEACH MATERIALS

Sand dominated by quartz or silica makes up more than 50 percent of all beaches from the polar regions to the tropics. This is because quartz is the most resistant to erosion of the common rock-forming minerals. Indeed, quartz sands are so resistant that they may remain in the coastal zone for hundreds of thousands of years. The sediment for sandy beaches may be derived directly from the erosion of adjacent cliffs, but most beaches are built out of river sediment deposited onto the continental shelf.

The distribution of other beach materials is largely determined by climate. Muddy beaches are most prominent in low latitudes associated with tropical wet climates, where the breakdown of rock materials into their finest particles is most complete; rivers, in particular, carry large quantities of fine-grained material to the coastline. However, muddy beaches

John Lythgoe/Planet Earth Pictures

David George/Planet Earth Pictures

may form almost anywhere as long as wave energy is low and only the finest sediments carried. Gravel beaches are particularly prominent in temperate and high latitudes. Here, in cooler conditions, particularly where there is a great deal of frost action, rocks disintegrate rather than decompose, and the rivers carry gravel rather than finer materials. This is particularly so in areas that have experienced glaciation. Glaciers have brought down to the coastline large quantities of coarse material which has subsequently been sorted by wave action. During the last great ice age, more than 18,000 years ago, areas peripheral to those that were ice covered experienced what is termed a "periglacial climate". Although not permanently covered with ice, frost action was particularly severe, and production of gravel-sized material was extremely efficient. Much of the southern part of the British Isles experienced periglacial conditions, and the widespread occurrence of pebble or shingle beaches is a legacy of this.

Boulder beaches are also associated with glaciated and periglaciated hinterlands. Boulder beaches dominate islands such as Spitzbergen, the northern islands of Canada, and the Falkland Islands. However, rather surprisingly, boulder beaches may also be found in tropical latitudes, the result of the intense decomposition of rocks in warm, wet climates. Some rocks, such as granite, break down in stages and produce core stones, which are delivered to the coastline as relatively even-sized boulders.

Another type of beach material is produced by plants and animals. Where plant and animal life forms calcareous skeletons in large amounts, biogenic materials may dominate the beaches. The term "coquina" is given to beach materials composed almost entirely of shells. Although biogenic beaches can be found almost anywhere in the world, they are particularly common in tropical areas, where they are associated with coral reefs. Occasionally, island beaches may be made up almost completely of *Foraminifera*, mollusks, and coralline algae dependent on coral. These may be in the form of round plates about 1 centimeter (½ inch) or so in diameter, but probably better known are the star sands, which are made up of the small, pink, star-shaped shells of smaller foraminifera such as *Baculogypsina* species.

John Beatty/Oxford Scientific Films

Many tropical beaches are made up of coral and shell fragments, the skeletal remains of countless tiny organisms.

BEACHES THAT LAST FOREVER?

Beaches made of shell and coral are maintained by the high productivity of their adjacent shallow-water ecosystems. Individual fragments are maintained within the beach only for a few hundred years at the most. Similarly, where cliffs are receding rapidly the beach materials may be extremely young, that is, no more than a few years old.

However, other materials may have lives spanning geological time scales. Sand and boulders may last in the coastal zone through several periods of sea level fluctuations. In many parts of the world, beaches that were formed in the last interglacial period, 125,000 years ago, are now being reworked or incorporated in present beach deposits. Many large boulder beaches, in particular, accumulate in this way, and individual boulders may have been reincorporated into active beaches several times during the past million years. Some beach materials may be even older, having broken down from sandstone formed many millions of years ago.

ENERGY ABSORBERS

Beaches, as well as being formed by waves, also act as baffles against their energy, particularly in storms. Much of the energy of waves is dissipated when water percolates into beaches, particularly coarser ones; or, during storms, is used in shifting sediment from the beach to the offshore zone, from where it returns in better weather. Whilst beach width is largely determined by the availability of sediment and tidal range, the slope of a beach is constantly changing in response to wave conditions and is delicately adjusted to the amount of swash that runs up the beach as waves break. ●

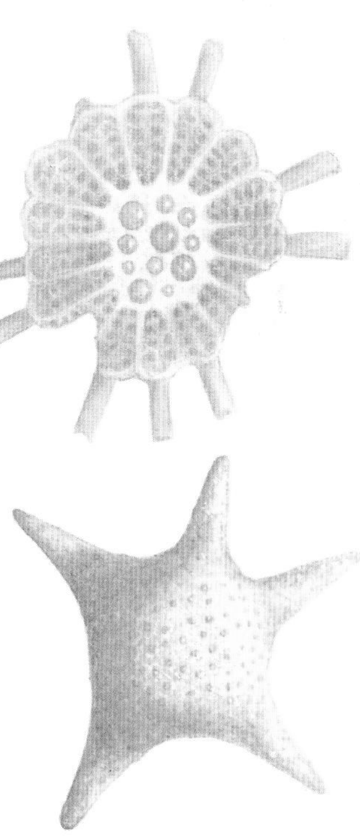

Two well-known Foraminifera *forms:* Baculogypsina *(top) and* Calcarina *(bottom).*

2 Continental Islands

DAVID HOPLEY

The term "continental island" covers a wide range of island types and sizes, from large continental fragments to small rocky outcrops, but all are characterized by their continental-type rocks and a history that shows they were previously connected to an adjacent continent.

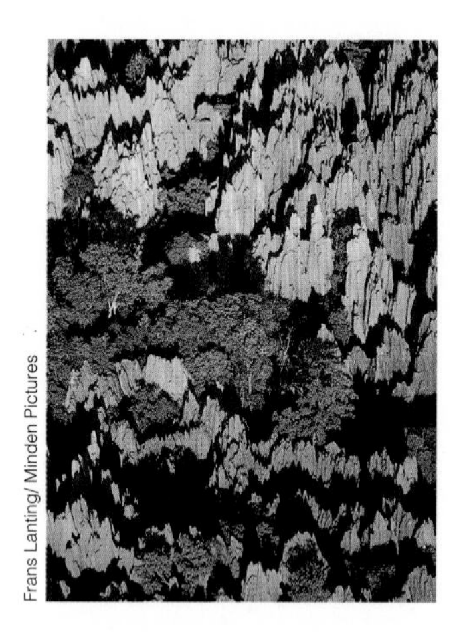

Frans Lanting/ Minden Pictures

Limestone pinnacles at Bemaraha, Madagascar. Madagascar is 587,040 square kilometers (226,660 square miles) and is one of the world's largest islands.

Continental Fragments

Continental rocks, dominated by silica and alumina (referred to as SIAL), are lighter than the silica magnesium rocks (SIMA) of oceanic crust. Because of this, continental fragments tend to be much more persistent on the Earth's surface, even where tectonic plates collide. It is always the oceanic crust which subsides beneath the continental rocks.

Throughout the history of the Earth, as the movement of new or subsiding oceanic crust has disturbed it, the continental crust has repeatedly fragmented and collided with other parts. In the past 200 million years, the continents Gondwana and Laurasia have broken up to form smaller drifting continents surrounded by continental shelves—submerged portions of the continental crust continually being built outward by sediments washed from the continents themselves. Some of these continental fragments have been small enough to qualify as islands. India was such a fragment before it collided with the Eurasian continent more than 70 million years ago. Other small continental fragments include Madagascar and Hispaniola (Haiti and the Dominican Republic). The continent of Australia is sometimes considered an island, although the continental shelf surrounding Australia extends as far north as Indonesia and includes the island of New Guinea as well as Tasmania in the south.

When a continental fragment lies close to a tectonic plate margin, it is likely to experience a violent evolution. Particularly where collision is taking place, followed by uplifting, volcanic and earthquake activity may change an island's features drastically, as has occurred, for example, in the islands of Indonesia and New Guinea. Where the continental shelf and its islands lie far from a plate margin, as do the islands off the southern shores of Australia, or northwest Europe, the most important processes affecting the islands' evolution may have been sea level changes.

In spite of their oceanic isolation, the New Zealand islands are continental and epitomize many of the varied features of large fragments of continental crust. The New Zealand continental fragment is far larger than the area currently outlined by the New Zealand

Andris Apse/Horizon

coastline. The complete fragment extends well to the south and east of South Island, and northwestward by way of the Lord Howe Rise, and by way of Norfolk Ridge to New Caledonia. More than 130 million years ago, New Zealand was part of Gondwana, joined to Australia, Antarctica, India, Africa, and South America. By about 100 million years ago, the New Zealand segment of the Indo-Australian Plate had separated from the other continental masses and become isolated from both Antarctica and Australia.

Not all New Zealand's rocks date back to the period before separation. The New Zealand continental crust fragment is large and flat, and for much of its time, sizeable areas of it have been beneath the sea collecting marine sediments, which are now exposed above sea level. The odyssey of this fragment of continental crust

Milford Sound, Fiordland National Park, New Zealand. Straddling two major tectonic plates, New Zealand is subject to vigorous geological activity, which has produced spectacular scenery, especially in the west of South Island.

brought it to a position where it now lies across the boundary of two of the world's major plates. Much of the western part of South Island and all of North Island is located on the Indo-Australian Plate, while the eastern and southern part of South Island lies on the Pacific Plate. This location has meant violent tectonic movement over the past few tens of millions of years; recently, it has produced the Southern Alps, which rise to more than 3,000 meters (9,800 feet). Active volcanism, particularly in North Island, and frequent earthquakes as the two plates move alongside each other indicate that mountain building is continuing.

RISING SEA LEVEL

Most continental islands, such as North and South islands of New Zealand, are still part of a major continental landmass, connected to it beneath the sea by the continental shelf. They have been isolated only in the past 15,000 years by the rise in sea level that occurred as the polar icecaps melted.

Usually these continental islands appear as clusters, groups, or archipelagos. This is hardly surprising as most have originated as the drowned remnants of former ranges and hills, and normally a line of coastal hills will extend seaward as a group of islands.

Continental islands form when rising sea levels flood coastal land around hills and mountains.

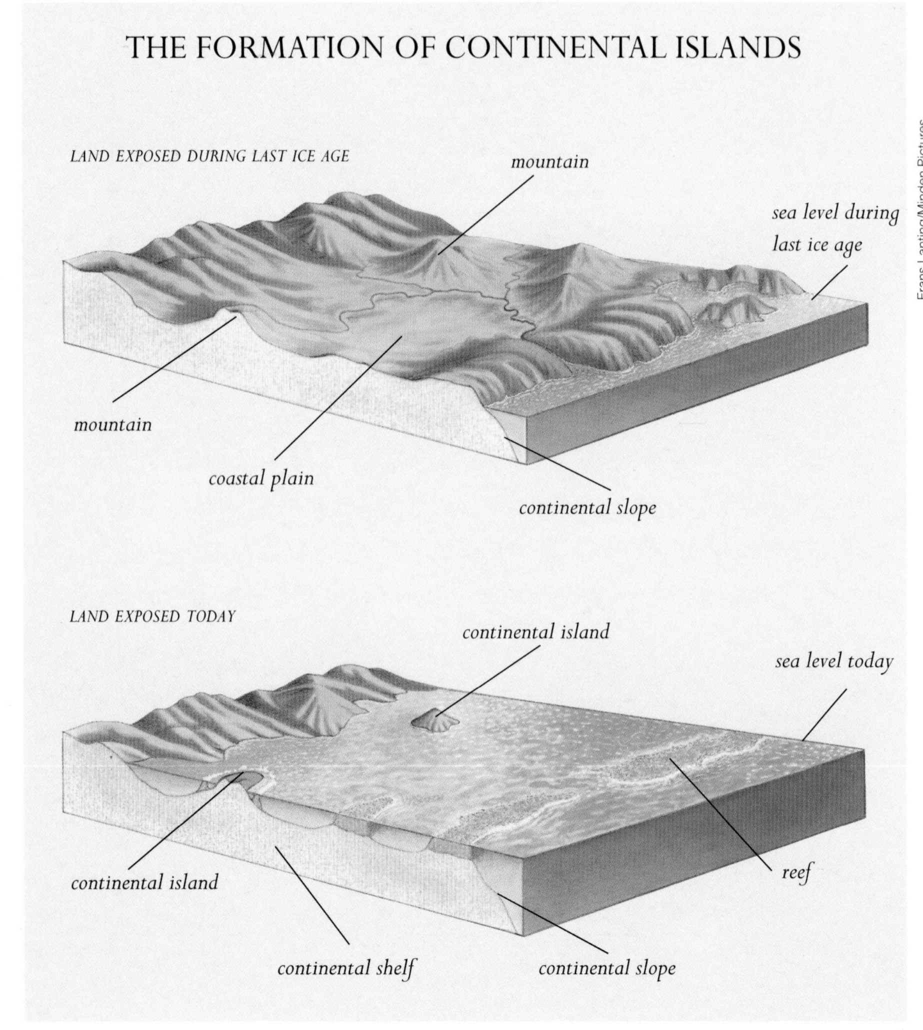

Frans Lanting/Minden Pictures

FORMING THE LAND

Tectonic activity may have the most dramatic effect on an island's topography, but other factors contribute to its character. An island's composite rock type and its history of erosion are particularly important. Different rocks have varying susceptibilities to erosion, and therefore produce distinctive landforms and islands. The importance of erosional history is seen best in islands affected by glacial erosion. As glaciers proceed down mountainous areas, they carve huge U-shaped troughs and steep ridges. On drowning, these produce highly indented coastlines, typified by deep fiords. The Lofoten Islands, the Hebrides, and many of the islands off the coast of Chile and western Canada and Alaska, including Vancouver Island, are examples of islands formed from the flooding of old glacial valleys. Lowland areas are also intricately carved by glaciers to produce irregular craggy outcrops. When the sea rises around these, they form the typical skerries such as

Vancouver Island on Canada's west coast is separated from the mainland by a maze of islets, channels, and fiords formed by the relatively recent flooding of glacial valleys.

those found off the coast of Finland or along the Skjaergard coast off Norway.

Thus, we can see a third factor at work in shaping continental islands—the climate to which the island has been exposed. This is important not only for the superficial appearance of vegetation, but also in determining which erosional processes sculpture the landforms. Besides glaciation, tropical weathering may be responsible for etching out spectacular peaks and domes, and producing thick layers of weathered material in areas of high relief. An island is, therefore, the product of its tectonic location, its rock type, and its erosional history.

Depositional Islands

The processes of weathering that change an island's landform also produce new islands. Depositional islands form out of the sediment from erosion, and the sand of coastal beaches and dunes. The common feature of depositional islands is the way in which they are molded by the sea. Such islands are often ephemeral and particularly susceptible to changes in weather patterns, rises in sea level, and erosion.

GLACIAL DEPOSITION

Coastal islands of glacially transported materials may exist where rising post-glacial sea levels have flooded around old terminal moraines. The seaward front of such islands will have been trimmed and straightened by wave action, with barrier beaches and lagoons being commonplace; on the inland side facing the mainland, the shores are likely to be strongly indented.

One of the most populated and well-known morainic islands is Long Island, New York, extending from Manhattan at the mouth of the Hudson River, 190 kilometers (120 miles) to the northeast. Three terminal moraines extend along the length of the island, and the island reaches 120 meters (400 feet) above sea level at its highest point. The sheltered

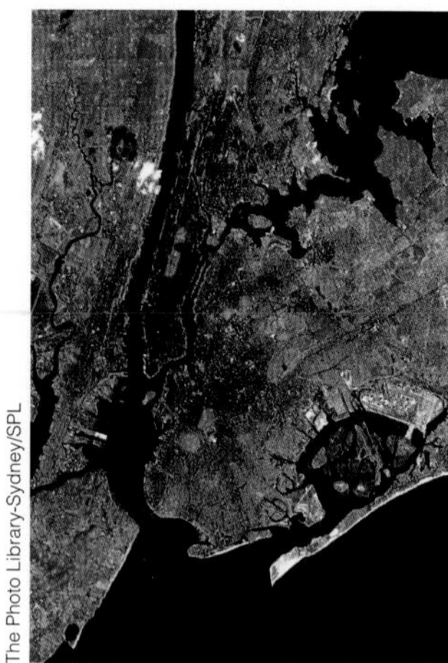
The Photo Library-Sydney/SPL

A Landsat infra-red image of New York and surrounding districts. Long Island is on the right and is connected to Staten Island, at bottom left, by the Narrows Bridge. The spit at bottom right is Rockaway Park, which partially encloses a cluster of islands in what is known as Jamaica Bay .

northern shore facing Long Island Sound has sandy bluffs interrupted by many bays and inlets, whereas the south is flanked by barrier beaches formed by the vigorous wave action of the open Atlantic. Behind the beaches lie salt marshes and mudflats.

Much of Long Islands's salt marshes have been reclaimed for urban use, and its southern beaches and parks make it one of the world's most popular playgrounds. Despite the encroachment of the city, agriculture remains important in the eastern part of the island.

The North Frisian Islands of Germany, close to the Danish border, have been formed largely from remnants of glacial deposits laid down during several ice ages, and they have only partly stabilized over outcrops of sedimentary rocks. The largest of these is Sylt, a hammer-shaped island, which clearly reflects the action of waves coming particularly from the north and west. The island covers an area of 99 square kilometers (38 square miles) and has a north–south length of nearly 40 kilometers (25 miles). The highest point is 52 meters (170 feet) above sea level; about 40 percent of the island risks flooding and lies below the high-tide level. The island's main ridge, formed from glacial moraine, has been eroded and the materials redistributed by waves to both north and south. In turn, the strong winds from the North Sea have picked up the finer sediments from the beaches and formed large dune sequences along both the northern and southern spits of Sylt.

Since 1855, Sylt has been a very popular seaside resort in northwest Germany, and what is geologically merely a redistribution of sediment during storms is now perceived as a major problem of coastal erosion. The situation is exacerbated by the fact that northwest Germany lies in a zone which is subsiding at a significant rate, and the open coastline of Sylt is retreating at a rate of about 1.5 meters (5 feet) per year. Not surprisingly, major erosion controls have been implemented, including sea walls and break-waters, and more recently, the use of massive concrete structures known as "tetrapods" at a number of strategic locations along the coast. However, a more effective measure has been found in artificially replenishing the beaches by pumping sand from offshore.

Eroded by wave action and land subsidence, the popular holiday island of Sylt on the coast of Germany has undergone some extensive erosion controls in a bid to conserve the diminishing coast.

THE ISLAND OF SYLT

List
Kampen
Wenningstedt
Westerland
SYLT
Keitum
Rantum
Morsum
Hörnum

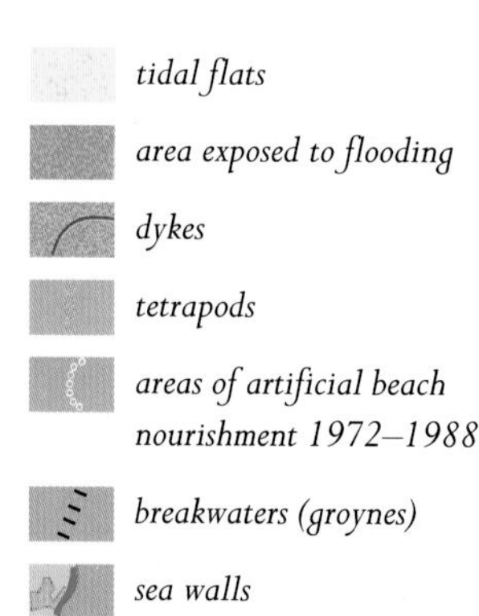

DELTAIC ISLANDS

Deltas form where a river enters the sea carrying more sediment than can be immediately redistributed by the wind and waves. Particularly where wave energy levels are low, a river's natural embankments, or levees, may extend seaward and divide several times as the river changes course. The most famous deltaic islands are those of Venice, but they occur in every delta.

This Mississippi river system carries between 1 and 1.5 million tonnes of sediment daily. Two percent of this load consists of fine sand; 98 percent is silt and clay. Within the deltaic environment, many types of islands can form. Each time the Mississippi floods, the river's mouth changes, as sediment is deposited, and new "crevasses" are formed, leaving fragments of the old levee banks and their associated marshes as isolated deltaic islands.

As a delta builds outward, the finest muds and clays are carried farthest into deeper water. Gradually, the coarser sediments, including sand and silt, which form the part of the delta above sea level, encroach seaward and form a cap over the mud. The weight of this cap on the very fluid muds can cause them to be squeezed like toothpaste, upward through the overlying formation, from depths as much as 1,000 meters (3,300 feet). If this happens in shallow water, small islands called "mudlumps" can form, the whole process taking only a few hours. Many mudlump islands are found in front of the Mississippi Delta.

Once a river changes course and its delta mouth has been abandoned, the fine sediments are quickly removed, and the sands redistributed into linear barrier islands. However, they are at the constant mercy of the wind and waves. As sediments are

Jean-Paul Ferrero/AUSCAPE

compressed beneath a delta, all its islands may be subjected to subsidence as well as erosion. The Chandeleur Islands, off the eastern side of the Mississippi Delta, form a moving bank 60 kilometers (37 miles) long. Surveys since 1855 have shown that these islands have gradually been driven to the west before easterly storm waves, and have moved as much as 1.5 kilometers (1 mile) in little over 100 years. A major hurricane in the area can move the islands hundreds of meters, and in extreme cases obliterate them completely. The Tiger and Trinity shoals are the remains of former islands off the Mississippi.

BARRIER ISLANDS

Longshore drift (the process by which waves approaching a coastline obliquely move sediment parallel to the shore) and the migration landward of large masses of sand laid down on the continental shelves during lower sea levels, can form offshore bars and spits, which gradually gain in size and form beach-barrier islands. These islands occur throughout the world, but particularly well-studied examples are the Frisian Islands of the North Sea and islands found along the coast of Texas, in the United States.

A feature of barrier islands is that they form long, straight coastlines behind which the sea, encroaching upon the land, forms a very indented inner coast. Individual barrier islands can be tens of kilometers long, and they are separated from each other by unstable tidal inlets. In the lee of barrier islands, both attached to the island and on the mainland shore, are extensive areas of salt marsh or mangrove. Since the postglacial rise in sea level, these vegetation communities have continually grown upward with the sea level, forming thick peat deposits. However, the

NASA/Science Photo Library

Fraser Island, off the coast of Queensland, Australia, is one of the largest sand islands in the world. It was formed from the consolidating effects of winds, tides, and ocean currents.

The formation of islands by siltation at the mouths of large slow rivers is effectively portrayed in this satellite image of the delta of the Mississippi River.

rise in sea level has also driven landward the main sand bodies of the islands. Thus, the sandy barriers overlie salt marsh peats, with peat frequently outcropping on the lower part of the ocean beaches.

As barrier islands are naturally retreating, any developments on them are environmentally suspect. Unfortunately, in the United States in 1968, the National Flood Insurance Program legally obligated the federal government to reimburse flood victims for damage and loss. Heavy subsidies of up to 80 percent made the NFIP very attractive to coastal property owners, and acted as an incentive to developers, particularly on the barrier islands. The insurance program was cut in the 1980s as a cost-saving measure. Nonetheless, during the 20-year period it was operative, substantial developments took place on barrier islands around the entire United States coastline.

Superb examples of barrier islands are the massive dune islands of southeastern Queensland, Australia. Here, a high energy swell coastline (which has allowed the sweeping of all sediment from the continental shelf since the post-glacial rise in sea level), a constant supply of sand from the rivers draining an eroding hinterland, and wind strengths sufficient to move sand from the beach for most of the year, combine to raise the islands into massive dunes more than 300 meters (980 feet) high. Moreton Bay, off Brisbane, is enclosed by the great dune and barrier islands of North and South Stradbroke, Moreton, and Bribie. Because of the subtropical location, mangroves grow in their lee, rather than the salt marsh found behind the temperate barriers of eastern United States.

One of the world's largest sand islands is Fraser Island, 120 kilometers (74 miles) long, 22 kilometers (14 miles) wide, and reaching a height of 240 meters (790 feet), off the Australian coast just south of the Tropic of Capricorn. Apart from four small outcrops, this island is made up of beach and dune materials accumulated over more than two million years of oscillating sea levels. Vegetation, including rainforest and ferns, covers much of the island.

Over long periods of time, after sand-dunes have been stabilized, organic matter is leached down to lower substrates where it helps to cement the sand together, forming an impermeable layer over which groundwater can rest. This produces the remarkable phenomenon of perched water tables and, when the ground surface intercepts the water table, perched lakes, which are entirely surrounded and underlain by sand. There are 40 such lakes on Fraser Island, including the largest in the world (Lake Boemingen) at 200 hectares (490 acres), and the highest (the Boomerangs) at 130 meters (430 feet) above sea level. ■

Joel Sartore

Where ocean currents move parallel to the shore, sediments from rivers may be swept into long narrow heaps, forming barrier islands along the coast. This barrier islands lies at the edge of a salt marsh in Louisiana.

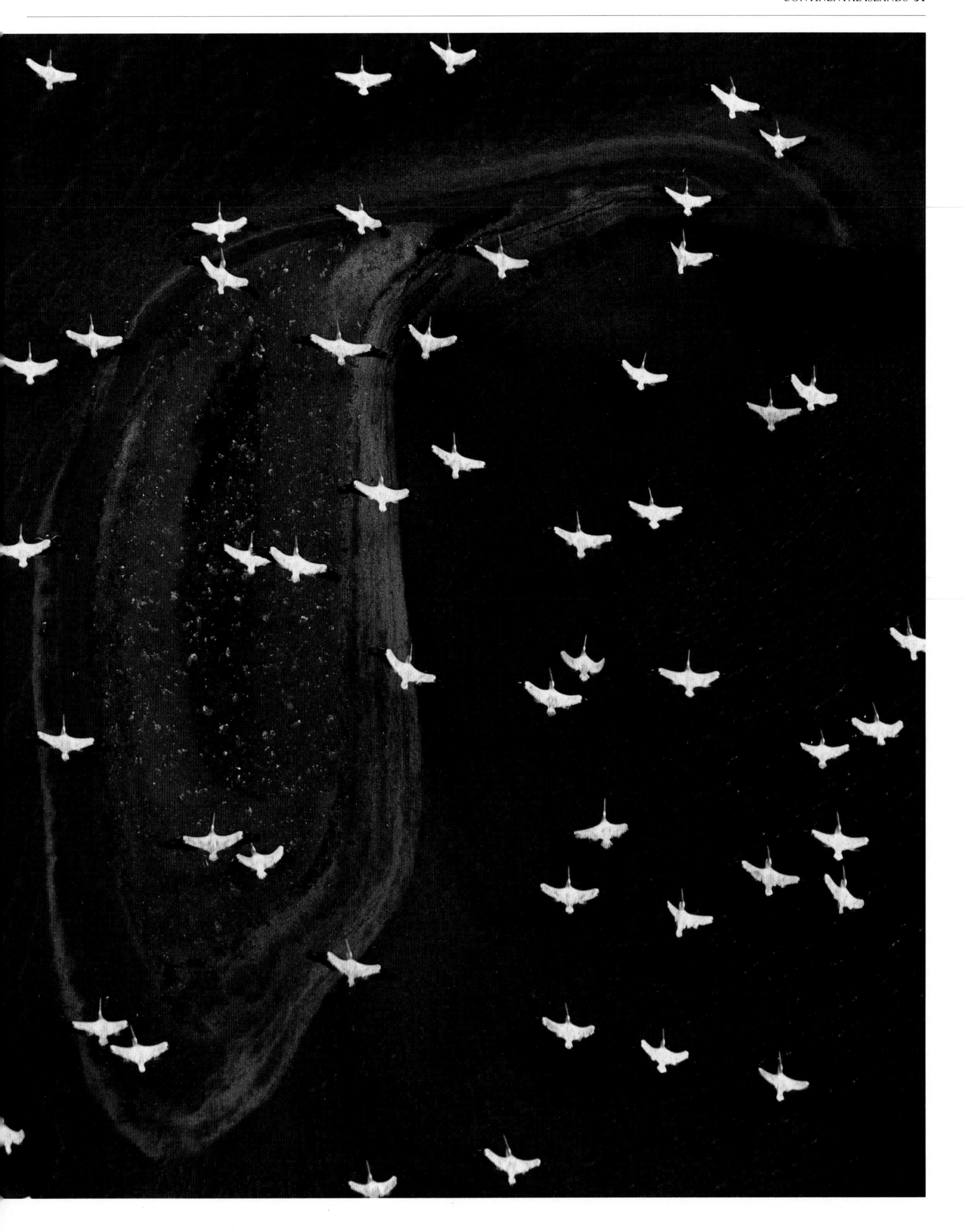

AUSTRALIA: THE ISLAND CONTINENT

PETER G. FLOOD

About 100 million years ago, fracturing occurred within the major southern landmass, Gondwana, and the Australian continental fragment began to break away from the Antarctic fragment. Australia (and New Guinea) moved northward, at a rate of approximately 5 centimeters (2 inches) per year. By about 45 million years ago contact was severed, and since then, the island continent, with its living cargo of plants and animals, has been like a Great Southern Ark.

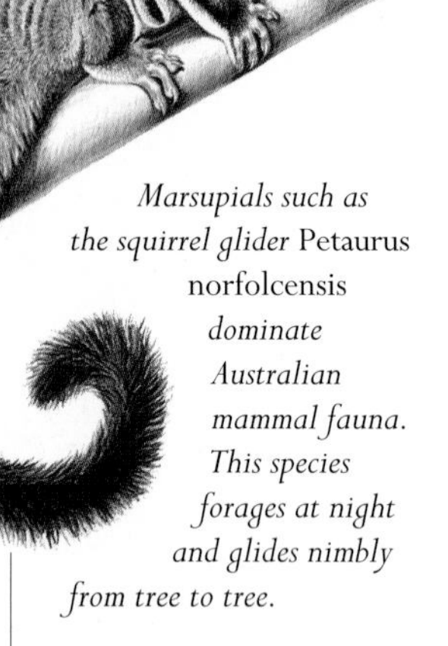

Marsupials such as the squirrel glider Petaurus norfolcensis *dominate Australian mammal fauna. This species forages at night and glides nimbly from tree to tree.*

The southern cassowary Casuarius casuarius *of New Guinea and north-eastern Australia belongs to one of nearly 20 entire bird families found nowhere else in the world.*

During this period, vastly different climatic conditions have prevailed. Thirty million years ago, Australia lay largely within the belt of moisture-laden westerly winds. At that time rainforests cloaked the eastern and southern parts of the continent, and open grasslands existed in the arid center. The world's temperatures were a few degrees warmer, and the seasons less marked. As the island continent continued northward, it passed through different climatic zones, eventually becoming warmer and drier. Today, much of Australia lies between 20°S and 30°S, that is, in the dry subtropical belt, where deserts form from the dry air that has already risen, cooled, and precipitated most of its moisture near the equator.

Gary Bell/Planet Earth Pictures

The separation of the continental fragments coincided with the demise of the dinosaurs and, in Australia, the stage was set for a new burst of evolutionary change by the now-isolated class of mammals which expanded to fill the ecological niche. Island biota unable to adapt to the changed climate died out. Mammals, including kangaroos, wallabies, bandicoots, numbats, koalas, possums, and platypuses, flourished, reaching their greatest variety with the spread of grasslands.

Mary Clay/Planet Earth Pictures

About 10 million years ago, the drifting island continent collided with the Pacific Plate in the north. This collision produced the spine of the New Guinea highlands, and brought the Australian continental landmass closer to the Eurasian Plate and the numerous islands of the Sunda Shelf. It also brought closer the separate biota of the two different plates. Migration in both directions was originally difficult because of the water barriers between the islands. As the colliding continued, the inter-island distance decreased, and island-hopping provided a "land bridge" between Australia and Southeast Asia for a limited interchange of biota.

The zone of contact, located in the Malay Archipelago, is known as Wallace's Line.

The drift of the island continent toward the equator has controlled the development of the Great Barrier Reef, which lies along the continental margin of northeastern Australia. About 17 million years ago, the northern tip of Australia came to lie fully within the latitudinal zone that suits luxuriant reef growth. Coral larvae from the already flourishing central Pacific reefs were able to colonize the rock substrates of the continental shelf. Between 14 and 12 million years ago, the reef ecosystems declined because of a drop in the sea water temperature related to the development of polar ice caps, and because increased rates of subsidence of the continental margin were drowning the reefs. Some reefs were able to keep pace with the sea level changes, and they have persisted until the present. Thus, the major form of the reefs had developed long before the present period of reef-top growth commenced.

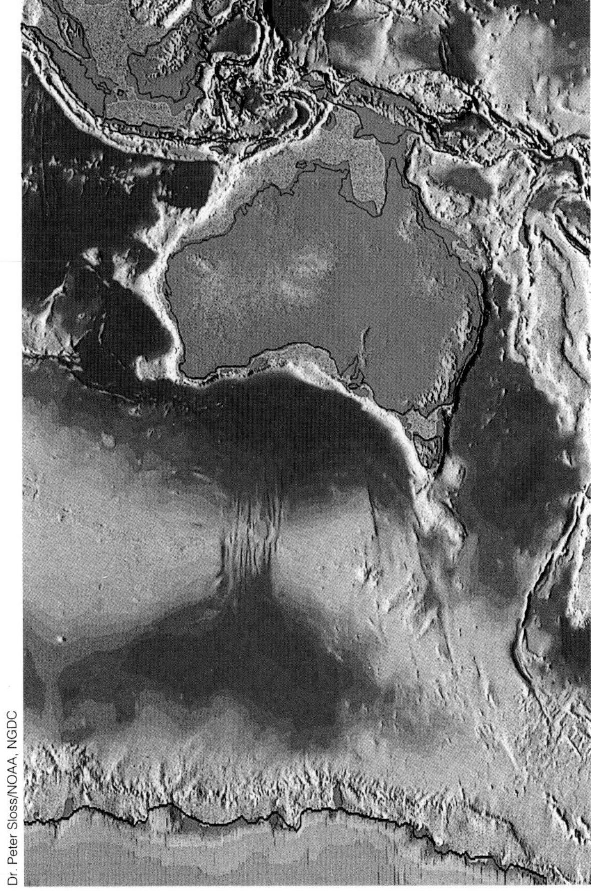

Dr. Peter Sloss/NOAA, NGDC

In its multimillion-year journey northward from Antarctica, Australia lies presently in the dry zone between equatorial and temperate regions. Its essentially low, level, and arid profile are starkly outlined in this colored digital image.

Center. *Australia's southern coastline at Port Campbell National Park, Victoria. Ceaselessly battered by Southern Ocean winter storms, the extensive sandstone coastal plateau has been eroded by wave action to form dramatic cliffs, arches, and stacks.*

The periods of glacial advance 10 million years ago saw the level of the sea fall dramatically by some 150 meters (500 feet), and the sea barriers to migration diminish. Around this time, the placentals (rodents) arrived in Australia from Southeast Asia. In the last 2 million years, glacial advances resumed and the Arafura Sea receded. The continental shelf surrounding Australia became exposed, thus creating land bridges linking the mainland of Australia and New Guinea. Similarly, the Southeast Asia Peninsula extended to include much of Indonesia, and it was by way of these extensive, newly exposed land bridges that new biota, eventually including the genus *Homo*, appeared in Australia.

If the present direction and rate of movement of the Indo-Australian Plate continues for another 20 million years, the entire island continent will enter the tropical climatic zone, and will eventually become part of the Southeast Asia landmass. ●

Marooned for about 45 million years, Australian marsupials radiated to fill a wide range of habitats. The numbat Myrmecobius fasciatus *feeds on ants and termites, and it exploits a niche occupied elsewhere in the world by anteaters and pangolins.*

3 Islands of Fire

RICHARD S. FISKE AND ALEXANDER MALAHOFF

The Earth is a dramatic structure, undergoing constant change. Nowhere, perhaps, is this more apparent than in the birth and death of volcanic islands, which have their origin deep within our planet, where molten rock wells up and bursts through fissures to the Earth's crust. As the molten rock—magma—cools and solidifies, deposits of lava are left, slowly building up on the crust. Although most of the Earth's volcanic eruptions occur on the ocean floor, the lava deposits may grow so high that they break the ocean surface. Thus is an island born of fire.

From Volcanoes to Islands

The acceptance of the theory of plate tectonics and the process of sea-floor spreading has, in turn, given us an understanding of volcanism. The most active volcanoes lie in belts which coincide with mid-ocean ridges and zones of fracture and collision between the Earth's plates—that is, the plate boundaries.

The source of the original magma is located in the asthenosphere, the plastic, partially molten layer lying beneath the lithosphere, the solid, elastic surface of the Earth. This could be as shallow as the base of the lithosphere (60 kilometers/40 miles), or as deep as 250 kilometers (150 miles) from the surface. As magma moves upward through fissures and rifts created by the movement of the plates, it loses some of its gases and gains some chemical constituents from the neighboring rocks, thus turning into lava. When lava solidifies, it forms volcanic rocks of which there are three types: basalt, the most widespread volcanic rock; andesite, formed by the volcanoes of island arcs located behind oceanic trenches; and rhyolite, the most siliceous volcanic rock, also formed by island arc volcanoes. All three are composed of silicon oxide and the oxides of other metallic elements. The tectonic setting of the eruption and the composition of the volcano's rock determines whether a volcanic island is mid-ocean ridge, island arc, or hot spot.

Katia Krafft/EXPLORER-AUSCAPE

Fresh lava oozes from a fissure on the slopes of Kilauea, Hawaii. High in temperature, low in viscosity, these basaltic pahoehoe *lavas cool in distinctively ropey, smooth-skinned formations.*

Opposite. *One of nature's great spectacles. Kilauea is the most vigorous of Hawaii's active volcanoes. It spews molten lava in fountains up to 600 meters (2,000 feet) high.*

Mid-Ocean Ridge Volcanoes

The separation of neighboring plates at the mid-ocean spreading centers, or ridges, leads to a reduction of pressure and an upwelling of magma from the asthenosphere to the surface. Slow-spreading sections of the mid-ocean ridge system, such as those found along the Mid-Atlantic Ridge, with spreading rates between 2 and 7 centimeters (1 and 3 inches) per year, may experience submarine volcanism at any given site only once every 10,000 years. On the other hand, fast-spreading ridge segments found along the East Pacific Rise, with spreading rates between 10 and 18 centimeters (4 and 7 inches) per year, may experience submarine volcanism at any given place once every few years.

Where magma extrudes along the fissures of the mid-ocean rift valleys, the asthenosphere has penetrated the surface of the Earth. At this line of active volcanism, there is no oceanic crust and no lithosphere. Basaltic magma, produced in mid-ocean ridges and hot spot volcanoes at high temperatures, is highly mobile and not very explosive. As the new crust forms after the lava has cooled, the lithosphere starts to develop. As the lithosphere cools and moves away from the ridge crest, it thickens and sinks, and after more than 90 million years attains a thickness of 60 kilometers (37 miles).

Mid-ocean ridge volcanoes, fed by a persistent, powerful magmatic plume, may form very large volcanoes that grow above the sea surface to create islands. Iceland, the largest mid-oceanic island in the world, represents such a plume-driven volcano located astride the northern end of the Mid-Atlantic Ridge, and is an excellent example of how sea-floor spreading works on the ocean floor. Iceland is in a

A contemporary depiction of the explosive eruption which destroyed the city of St. Pierre on Martinique in 1902 and left only two survivors.

Three geological processes form most volcanoes: island arc volcanoes may occur where one ocean plate collides with another; mid-ocean ridge volcanoes are associated with sea-floor spreading; and hot spot volcanoes form from persistent magma plumes that rise from deep within the asthenosphere.

state of east–west tension, one half located on the North American Plate moving westward, the other half located on the Eurasian Plate moving eastward. Tension cracks develop, and magma flows up the cracks and erupts as basaltic lava on the surface. The volcanic activity has created some striking landforms. In one single eruption in 1783 the Laki Fissure, a fissure 32 kilometers (20 miles) long, opened up and erupted 12 cubic kilometers (3½ cubic miles) of lava. The magma solidified in the fissure and formed a vertical wall of dense basalt, known as a dike.

ISLAND ARC VOLCANOES

Island arc volcanoes—among them the Aleutians, the Antilles, and the islands of the Indonesian Archipelago—are the result of a collision between two plates in mid-ocean. There are about 22 long, curved chains of these islands in the world. They show remarkably similar features, including narrow, deep-sea trenches along the convex side of each of the arcs, and chains of volcanoes of particularly explosive activity.

At the line of convergence, the thinner, denser oceanic plate slides under the other to form a deep trench. In this process of subduction, the oceanic crust is consumed and remelted at depths of 700 kilometers (430 miles) beneath the surface. Friction between the downward-moving (subducting) crust and the plate margin leads to heating and partial melting of the subducting plate at a depth of about 200 kilometers (125 miles) beneath the surface. The mixture of remelted basalt and melted oceanic sediments overlying the subducted crust forms a siliceous, gassy magma that migrates to the surface and erupts in the form of andesitic and rhyolitic lavas.

Andesitic and rhyolitic lavas build cone-shaped, steep-sloped volcanic islands such as those of the Izu-Bonin islands located south of Japan, or the Aleutian Island chain. In the Andes (South America), such volcanic cones reach heights of 6,800 meters (22,300 feet). In the Mediterranean Sea, Mount Etna (Sicily) rises above sea level as a volcanically active cone, as does Mount Stromboli.

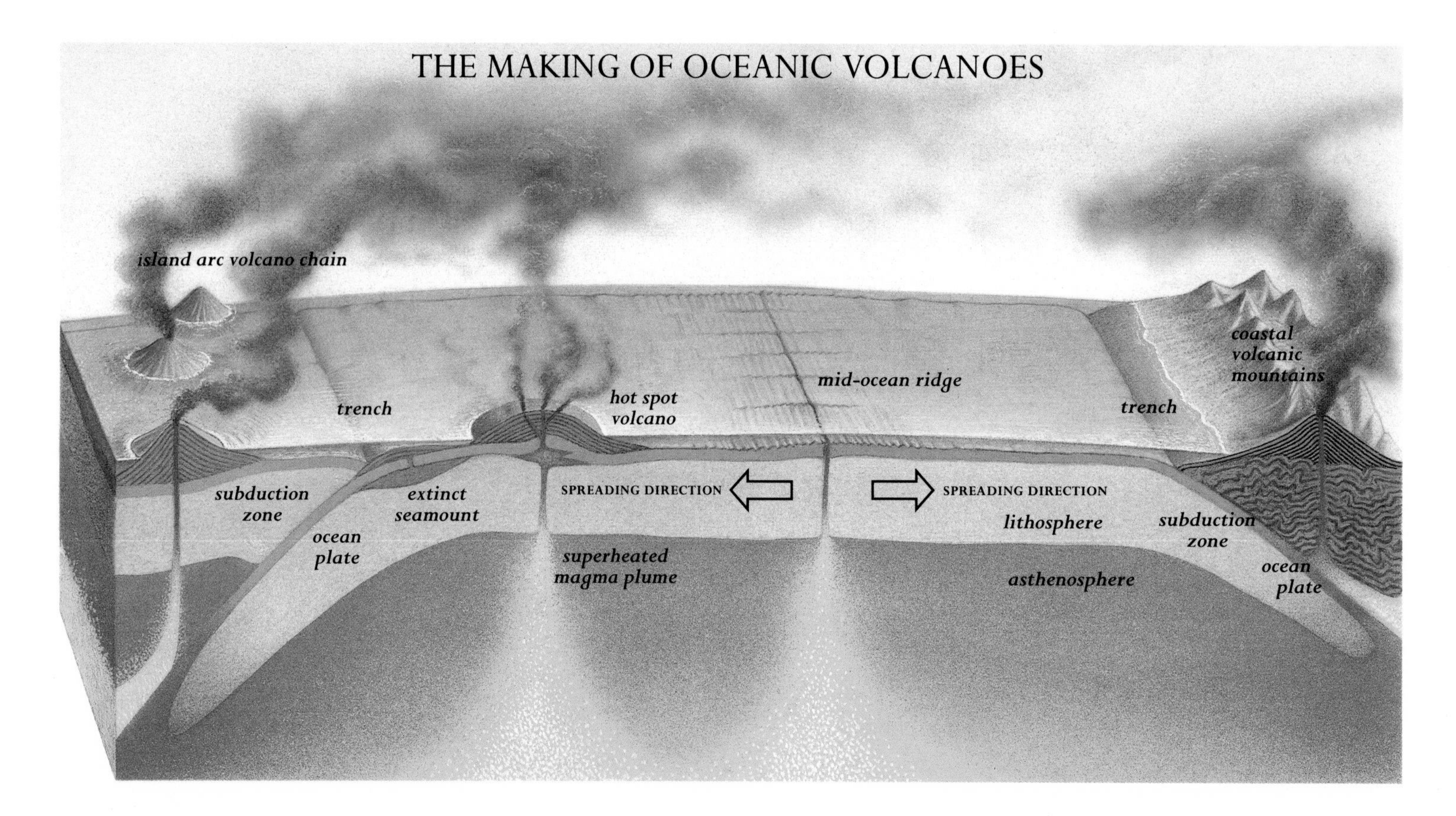

Katia Krafft/AUSCAPE

The cones of these volcanoes are frequently severely reduced in size, and may even disappear after steam-driven or phreatic explosions. The highly siliceous and gassy rhyolitic lava, created at a lower temperature than basaltic lava, produces volcanism that may be highly explosive, such as that of the 1980 Mount St. Helens eruption in the United States.

Rhyolites can produce eruptions of incandescent clouds made up of glowing high-temperature particles bathed in steam, and moving at more than 100 kilometers (60 miles) per hour. An example of this kind of volcanic force was the explosive eruption of Mont Pelée in Martinique in 1902, which destroyed the town of St. Pierre and killed about 30,000 inhabitants.

Giant explosions of these volcanoes have destroyed whole island civilizations. About 1470 BC the near-shore island volcano of Thira (Santorini) exploded in a steam and ash eruption, destroying the Minoan civilization then flourishing on the island. In 1815 the volcano Tambora on the Indonesian island of Sumbawa erupted and ejected 150 cubic kilometers (44 cubic miles) of pumice and ash into the atmosphere, which led to an endless winter in North America and Europe during the year of 1816.

More recently, in 1883, the Indonesian volcanic island of Krakatoa was destroyed in a violent eruption which killed 36,000 people. Violent mixing of gassy

Photographed from the air, Mount Augustine in Alaska shows the classic conical form of a volcano, gradually built up from layers of lava, ash, and other materials ejected by a succession of violent eruptions.

KINDS OF MID-OCEAN RIDGE VOLCANISM

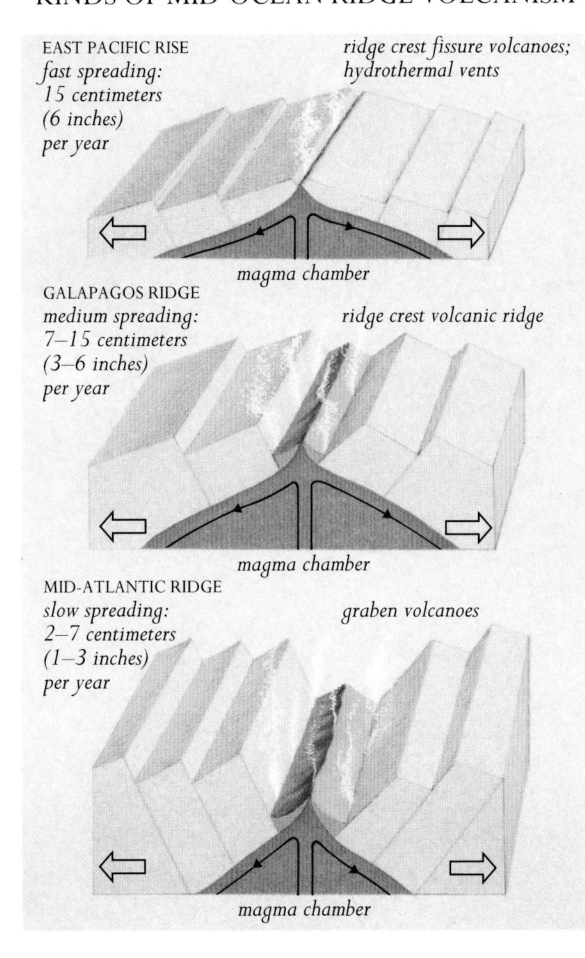

These ridge axis cross-sections depict three styles of mid-ocean ridge volcanism and faulting. The arrows indicate magma motion in the chamber and the direction of sea-floor spreading.

Krakatoa's explosion was heard more than 4,500 kilometers (2,700 miles) away and the area of ashfall was approximately 2,000 kilometers (1,250 miles). The fine volcanic dust served as a filter for solar radiation, lowering temperatures around the world for a year after the eruption.

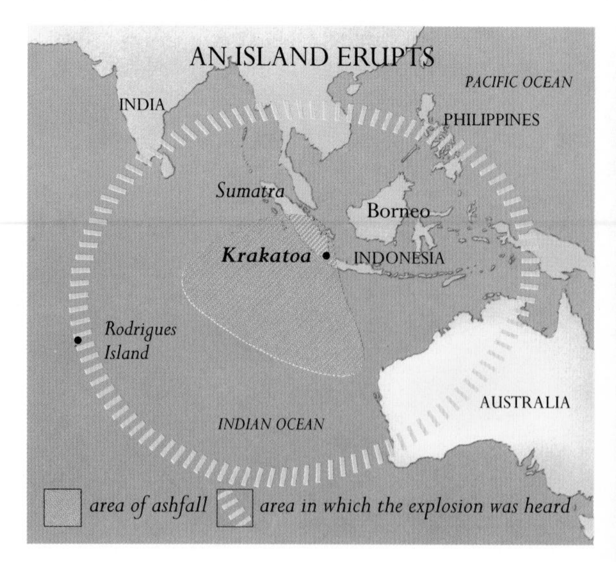

andesitic and basaltic magmas and sea water in the magma chamber of Krakatoa probably led to the destructive ash eruption. Similar near-shore island arc volcanoes with andesitic or rhyolitic magmas are active in the Mediterranean Sea, in the eastern Indian Ocean, and around all the margins of the Pacific Ocean.

HOT SPOT VOLCANOES

Thousands of islands and seamounts within the Pacific Basin and the Indian and Atlantic oceans were formed by the process of hot spot volcanism. Some hot spots have been persistent sources of submarine volcanism for as long as 70 million years and, because of the north and northwestward motion of the Pacific Plate, have built long island chains in the Pacific Ocean: the Cook, Austral, Tuamotu, Marshall, and the Hawaii–Emperor island chains.

A hot spot is a fixed magma source that lies deep in the Earth's mantle, well below the moving lithosphere. Why the hot spot exists in a particular place in space and time is not known. The localized heating of the hot spot is probably driven by the decay of the radioactive isotopes of uranium, potassium, and thorium in the Earth's mantle. The mantle is hotter, and therefore less dense, than its surroundings, which already are at a temperature of 1,800°C (3,270°F). The hotter, less dense material begins to rise in a similar way to a hot air balloon.

The upward rise of the magma is not continuous, but appears to occur in episodes, with the heated bubbles of magma rising in a plume to the base of the lithosphere. The lithosphere (and the asthenosphere) is in motion: in the Pacific, for example, it is moving in a northwesterly direction at a rate of 10 centimeters (4 inches) per year. Following the movement of the lithosphere, the path of the upward-moving magma curves northwestward from its source deep in the mantle and erupts initially as a submarine volcano, and later as an island volcano above the water's surface. In the case of the Hawaiian Islands, therefore, there is an increase in age northwestward away from the current hot spot volcanic activity at the Loihi submarine volcano, along the trajectory of motion of the lithosphere. Midway Island was formed by submarine volcanism 28 million years ago at the

Dorian Weisel/Planet Earth Pictures

geographic site of present-day Loihi. Since its formation, the island has moved on the lithosphere in a northwesterly direction for a distance of nearly 2,600 kilometers (1,600 miles). Meiji Seamount, located 2,500 kilometers (1,550 miles) north of Midway, is 72 million years old. This data tells us that between 70 and 28 million years before the present, the Pacific Plate was moving north at a rate of 6 centimeters (2½ inches) per year, or 60 kilometers (37 miles) per million years. Some 28 million years

A curtain of fire—the result of lava erupting along a fissure on the slopes of Kilauea.

Krafft/EXPLORER-AUSCAPE

ago, the direction of movement of the Pacific Plate changed to northwest, and the rate to almost 10 centimeters (4 inches) per year. This direction and rate have persisted to the present day.

Since the individual islands along the chain are separated by deep water, it stands to reason that the magma supply from the hot spot sources has waxed and waned over time. At present we cannot explain this phenomenon. As the island volcanoes moved away from their hot spot and lost their magma source and supply, the volcanoes became extinct, and began to sink and erode: Midway has eroded to sea level. Many of the others, such as Meiji, have sunk below sea level, to form flat-topped seamounts or guyots.

The Hawaiian volcanoes have been built up out of frequent eruptions of massive, free-flowing basaltic lava. (The more violent eruptions that result in ash cones are infrequent.) Basaltic eruptions of this type usually build up a shield-shaped volcano, rather than the cone shapes of island arcs. The Hawaiian shield volcanoes form broad islands that rise more than 10 kilometers (6 miles) above the ocean floor. Above sea level, basaltic mid-oceanic volcanoes, such as Mauna Loa and Kilauea, have characteristically mobile lava eruptions, which move at speeds of up to 10 kilometers (6 miles) per hour. With few explosive components, they appear as long incandescent rivers. These cool into ropy *pahoehoe* forms or move slowly downslope as *a'a* (block) lavas.

The Galapagos Islands are also an example of shield-shaped islands. Other islands formed by hot spot activity beneath the ocean plate include the Canary and Azores islands, Cape Verde, Tristan da Cunha, and St. Helena in the Atlantic, and the Tahitian, Marquesas, Easter, and Juan Fernández islands in the South Pacific.

Erosion of these high islands has produced the spectacular and craggy mountainous scenery typified by Tahiti and Oahu in the Hawaiian Islands.

THE VOLCANOES OF HAWAII

The Hawaiian Volcano Observatory is located on Kilauea, the most active of the Hawaiian hot spot volcanoes. There are three active volcanoes and one dormant volcano associated with the Hawaiian hot spot, all located within a circle, 100 kilometers (60 miles) in diameter. The three active volcanoes are Mauna Loa, Kilauea, and the Loihi submarine volcano; the dormant volcano is Hualalai. Mauna Loa, Kilauea, and Loihi apparently all have separate conduits along which magma moves up from the asthenosphere through the solid lithosphere to erupt as basaltic lava at the calderas and craters, and along the fissures of these volcanoes.

Both Mauna Loa and Kilauea have grown above prominent rift zones; their summits are marked by distinct calderas, where the top of the mountain has subsided. Mauna Loa (the highest volcano on the surface of the Earth) was active once every 4 or 5 years until 1950, and is now active every 10 to 25 years, with eruptive phases lasting for up to 18 months. During the past 150 years, lava poured out in volumes of up to 440 million cubic meters (15,530 million cubic feet) per eruption, an average rate of 21 million cubic meters (741 million cubic feet) per year. During this period, Kilauea erupted once every few years, with an average of 10 million cubic meters (353 million cubic feet) of lava per year. One of the most voluminous eruptions recorded recently has been that of Kilauea—a continuous eruption which started in 1983, producing an average of 150 million cubic meters (5,300 million cubic feet) of lava per year. In 1984, 230 million cubic meters (8,120 million cubic feet) of lava also erupted on Mauna Loa during a three-week period.

THE GROWTH OF LOIHI

Still to break the surface of the water, the submarine volcano Loihi, meaning "long one" in Hawaiian, has grown on a fissure system oriented north–south-to-southeast, approximately following the path of the hot spot trace on the ocean floor.

Active volcanic extrusion is taking place at the summit of Loihi in a water depth of 960 meters (3,150 feet), and at the tip of the southern rift, 21 kilometers (13 miles) southwest of the summit, in a water depth of 5,200 meters (17,000 feet). At the southern end of the rift, the 90-million-year-old oceanic crust upon which Loihi sits is fissured. Fresh pillow lava (formed as new outbreaks of incandescent lava burst out of the "pillows" already formed by the chilling of the water) and ropy *pahoehoe* lavas have poured out of this fissure, covering the surrounding sediments. Northward of this leading edge of the rift, pillow lava cones 100 meters (330 feet) high have been built along the rift by active volcanism. These lavas are fresh with just a dusting of ocean sediments covering them. The cones are mechanically unstable and mark the sites of frequent downslope slumping of the freshly erupted basalt.

The mixture of talus (debris) and pillow lavas that has built the edifice of Loihi is very porous, and sea water circulates freely through the rock that covers the slopes of the submarine volcano. The interior of the volcano is marked by vertical dikes filled with hot rock or liquid magma. Sea water interacts with the heated dikes, and is in turn heated up to 400°C (750°F),

Katia Krafft/EXPLORER-AUSCAPE

Little by little, grasses and ferns recolonize lava flows.

Opposite. *A dramatic fire fountain on Bali Batur, Indonesia, the result of a highly explosive island arc eruption.*

An extinct, eroding giant—the crater of Haleakula on the island of Maui. Maui has moved away from the hot spot that fuels the active Hawaiian volcanoes.

Comstock Photo Library

Kilauea is one of the most studied volcanoes in the world. Because its eruptions are relatively quiet outpourings of fluid lava, the Hawaiian Volcano Observatory has survived since 1912, perched atop the summit and providing a base from which the volcanic activity can be studied at close range.

Paul Chesley/Photographers Aspen

Fiery molten lava seeps slowly into the sea surrounding Hawaii causing huge clouds of steam to rise from the water.

G. Brad Lewis/Douglas Peebles

leaching the surrounding basalt of minerals and then migrating upward in the form of mineral-laden hot water. The summit of Loihi is characterized by pit craters, and by active hydrothermal vents, fissures through which the sea water exits. When this hot fluid reaches the surface and comes in contact with the cold sea water, the shock cooling of the hydrothermal fluid crystallizes the minerals in solution, forming chimneys and mounds of zinc, iron, and copper sulfides. Exotic life, in the form of tube worms, clams, and crabs feeding on the bacteria in the heated water, cluster around the hydrothermal vents. From observation of these vents, we now understand that metal sulfide deposits on land, such as those on the islands of the western Pacific, were originally formed around the hydrothermal vents of the ocean floor.

ALEXANDER MALAHOFF

Unstable Oceanic Volcanoes

In recent years geologists have learned that large oceanic volcanoes, such as hot spot volcanoes, have a far more eventful history than simple growth followed by subsidence. The important missing chapter in this history is "gravitational collapse"—the wholesale destruction of large parts of oceanic volcanoes. In a nutshell, volcanic edifices, built mostly of lava flows and loose fragmental material, are notoriously unstable. Not only do they fall apart because of gravitational instability, they tend to be pushed apart by pressures of magma within them. The combined effect has produced some of the most gigantic landslides in the world.

Huge landslides have been recognized at dozens of oceanic volcanoes, but they have been best documented in Hawaii. This may seem surprising, because the volcanoes of Hawaii have very gentle slopes and, therefore, would not seem susceptible to landslides. However, studies on the Hawaiian Islands and on the nearby sea floor have shown that these volcanoes are literally falling apart. Instabilities have been recognized in mature volcanoes, such as Koolau (Oahu Island), East Molokai, and Mauna Loa; in actively growing volcanoes such as Kilauea; and in youthful volcanoes such as Loihi, that have not even grown above sea level to form islands.

The pillow lava and related debris of which most of Loihi is built is apparently gravitationally unstable even while it is below sea level. Loihi erupts a few times each decade, and it is estimated that its summit will grow above sea level to form the newest of the Hawaiian Islands in 10,000 to 100,000 years. Obviously, the upward growth of Loihi will be controlled by the offsetting effects of eruptions and landsliding.

UNCERTAIN FUTURE

Kilauea, on the Big Island of Hawaii, is one of the world's most active volcanoes, and is in what is called the "shield building" phase of its life. Kilauea has been the site of dozens of eruptions since 1952, and has been erupting almost continuously since January 1983. Despite its rapid growth, and almost paradoxically, Kilauea is in the process of falling apart because of landslides and large slumps. Much of the entire south flank of the volcano is being displaced toward the southern side of the island at very rapid rates. One surveying line, reaching from the relatively stable slopes of adjacent Mauna Loa to Kilauea's south flank, has extended more than 4.5 meters (16 feet) since 1970. Recent studies have shown that these large displacements are caused both by gravitational instability of the volcano itself, and by the wedging actions of magma bodies within the volcano that tend to push its entire flank seaward.

For purposes of comparison, the area of Kilauea's south flank above sea level is about 780 square kilometers (300 square miles), and the entire south flank, including its undersea extension, is about

INSIDE KILAUEA

Decompressing magma rises to Kilauea's summit reservoir by forcing its way through rock fractures. The molten rock erupts either through the summit or through the eastern flank via a horizontal conduit.

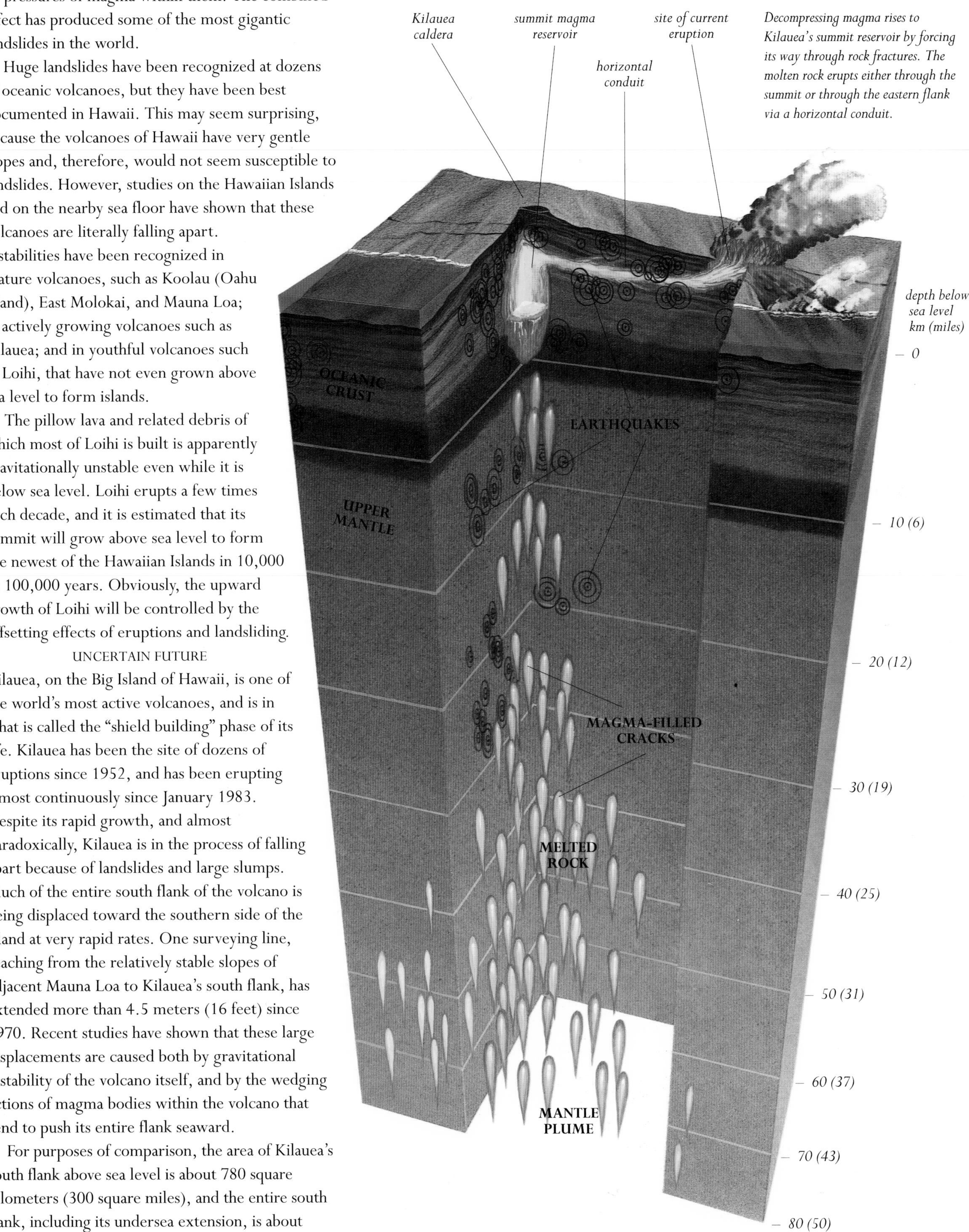

Protected by a heat-reflective metallic suit, a volcanologist maneuvers warily at the brink of an erupting crater. The temperature of molten lava typically ranges from 800°C to 1,100°C (1,500°F to 2,000°F).

3,890 square kilometers (1,500 square miles). Considering that this mobile flank averages about 5 kilometers (3 miles) in thickness, it becomes apparent that a volume of more than 16,000 cubic kilometers (4,000 cubic miles) of volcanic material is being moved seaward. It is unlikely that this segment of the volcano will slump into the sea in a single catastrophic event; instead, it is more likely to splinter off in smaller increments—perhaps in individual masses of a few cubic kilometers. In any case, it is clear that dramatic events are unfolding on the south flank of Kilauea.

LANDSLIDES OF THE PAST

Even more dramatic landslides have occurred from other Hawaiian volcanoes in the past. Recent side-scan sonar images have revealed jumbled masses of debris that extend outward from nearly all the Hawaiian Islands, as well as from the now-submerged volcanoes lying along the Hawaiian chain farther to the northwest. Two of the largest landslides, originating on the northeast side of the island of Oahu and on the north side of Molokai, have coalesced to form a hummocky debris field more than 225 kilometers (140 miles) long, containing individual blocks of these disrupted volcanoes larger than the entire island of Manhattan. The Alika debris slide, which originated when a large part of Mauna Loa collapsed 100,000 years ago, produced a tsunami (tidal wave) that ran up the side of the adjacent island of Lanai to a height of 280 meters (925 feet). If such an event were to occur today all coastal areas in the Hawaiian Islands would feel its effects; waves up to 30 meters (100 feet) high might roll in upon Honolulu. Clearly, we are dealing with catastrophes of biblical proportions.

Happily, the probability that such a catastrophic event will occur in our lifetime is extremely small. Nevertheless, the knowledge that giant landslides are important geologic processes should be kept in mind when considering Darwin's classic model for the growth and subsidence of oceanic volcanoes and the development of atolls. As usually presented, this model depicts an erupting oceanic volcano that retains its original geometry as it becomes quiescent and then subsides beneath the sea. The recent discoveries described here suggest instead that oceanic volcanoes commonly lose a significant part of their volume through the process of landsliding. One possible consequence of this is that newly developed fringing reefs, and even young atolls, could be partly destroyed by collapse of the volcanic edifice upon which they are growing. It is possible, therefore, that imperfections in present-day atolls, such as significant gaps in their normal circumference, might be due to earlier gravitational failures of their volcanic foundations. ■

RICHARD S. FISKE

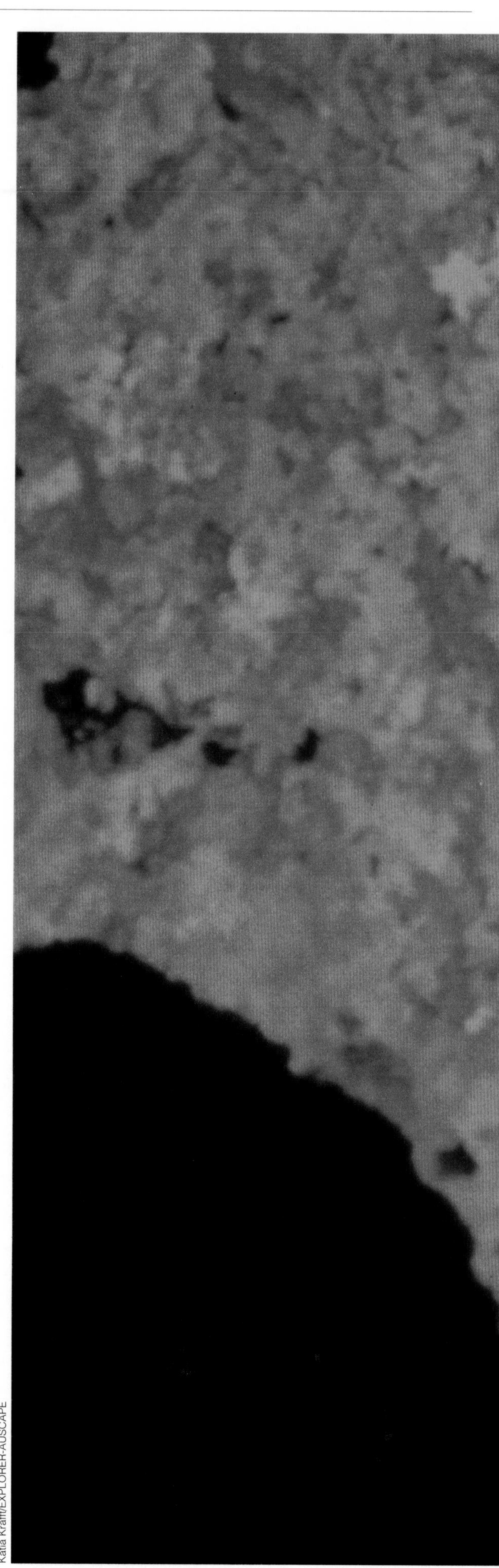
Katia Krafft/EXPLORER-AUSCAPE

4 The Formation of Coral Islands

RICHARD W. GRIGG

Coral islands represent some of nature's finest monuments, often characterized by unrivaled beauty and ecological complexity. Their majestic landscapes and underwater realms have captured the imaginations of writers, artists, explorers, scientists, and adventurers alike.

DARWIN'S MODEL

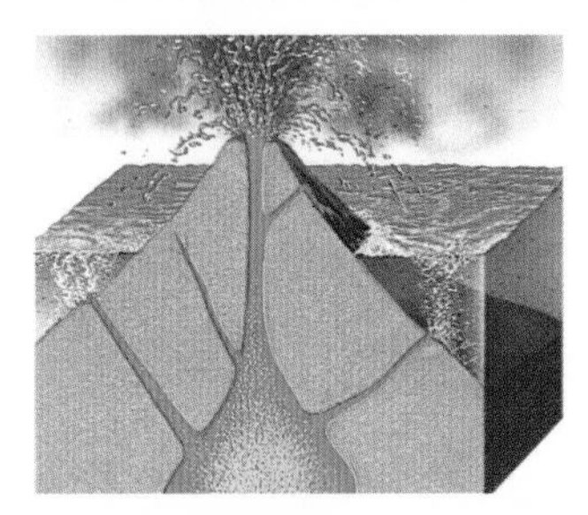

After a violent eruption a volcano emerges from the sea.

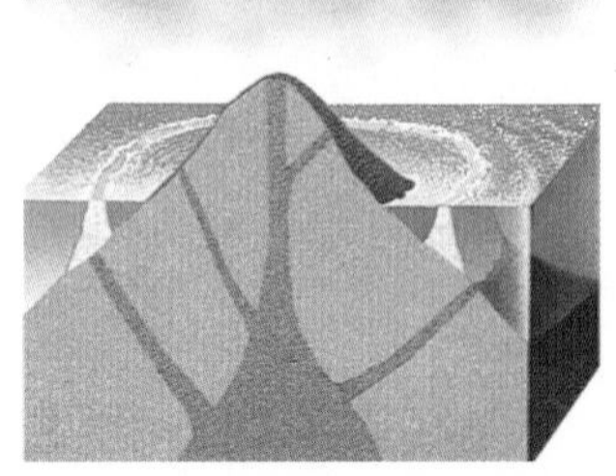

A fringing reef forms around the newly formed island.

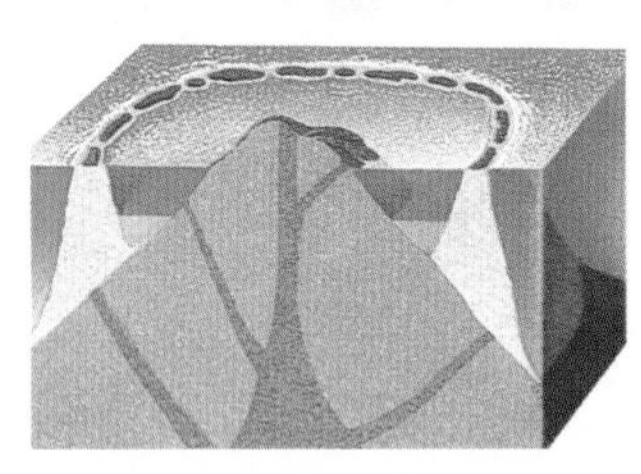

The island begins to subside and the reef develops into a barrier reef.

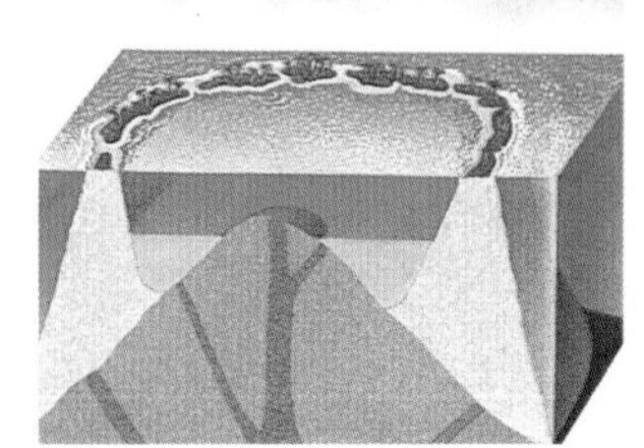

The island sinks into the sea leaving a circular or elliptical atoll.

Growing Islands

Most coral islands are the product of a rich synergy between geological and biological processes. With the help of symbiotic algae (plant cells) that live within most coral tissues, the coral animal secretes a skeleton of limestone, at a rate of about 1 centimeter (½ inch) per year. Over thousands, or even millions, of years these coral deposits build up to massive calcareous reefs sometimes hundreds of meters thick. Uplifted mountains over 300 meters (1,000 feet) high in Australia and China consist entirely of reef skeletons laid down during the Paleozoic era more than 250 million years ago.

Coral islands are usually described as either high or low islands. Most high coral islands are young volcanoes surrounded by reefs in various stages of development, although they may also consist of continental fragments, such as the Seychelle Islands, which are ringed by coral reefs. Low coral islands are nearly flat, and most are made up of the skeletons of coral and other calcareous reef organisms. Atolls are coral islands that consist of lagoons surrounded by coral islets and barrier reefs.

All coral islands begin with the upward growth of coral from undersea foundations. Because coral growth requires an ocean temperature generally above 20°C (68°F), coral islands are restricted to tropical latitudes between 30°N and 30°S. While coral islands can form on stable foundations, most coral atolls have grown upward from the summits of sinking drowned volcanoes. In 1835 Charles Darwin first postulated this as a hypothesis. He had climbed the slopes behind Papeete in Tahiti to view the island of Moorea, which is encircled by a lagoon and barrier reef. To transform such an island into an atoll, Darwin reasoned that if the islands were to sink slowly, the upward growth of the coral barrier reef would eventually produce an atoll.

Darwin's theory was confirmed more than a century later at Enewetak Atoll in the Marshall Islands where two holes drilled into the reef reached volcanic rock at depths of 1,267 meters (4,155 feet) and 1,406 meters (4,612 feet), respectively. Since coral reefs cannot thrive below 100 meters (330 feet) at the most, these results proved both that Enewetak Atoll had subsided more than 1,000 meters (3,300 feet) and that there had been constant upward growth during its long history of 49 million years.

Nicholas de Vore/Bruce Coleman Ltd.

FROM CORAL COLONY TO ATOLL

Coral reefs form around many different kinds of islands in tropical latitudes, not just volcanoes. For example, there are thousands of small, high islands in the Indonesian Archipelago, many of which provide shallow shelf environments for the richest coral reefs in the world. Over 80 genera and 500 species of reef-building corals exist in this region—the Indo-West Pacific province—which is the center of global diversity for many tropical plants and animals.

The smallest coral reefs are simply individual coral colonies that coalesce to form large heads, or patch reefs, several meters in diameter. Micro-atolls are a particular growth form of single coral colonies that result when upward growth is arrested by shallow sea level. When they reach sea level, the corals (usually Porites) continue

Captain Cook divided islands into "high"—towering volcanic peaks visible from afar—and "low", typified by this idyllic coral atoll in French Polynesia.

A coral predator: the crown-of-thorns starfish Acanthaster planci, *Seychelles. Widespread in the tropical Indian and Pacific oceans, this starfish periodically reaches plague proportions which can devastate a reef.*

to grow outward. Calcification continues around the outer margins of the colony while in the center coral tissues die off, leaving a saucer-like formation resembling an atoll in shape but only several meters in diameter.

As reefs enlarge, they often coalesce to form patch reefs of various sizes or, along the coastal margins, long veneers of coral that can stretch for many kilometers. These structures develop first into fringing reefs, which are usually shallow enough to cause ocean waves to break but are insufficiently developed to block their shoreward advance. When coral reefs are large enough to dissipate the energy of offshore waves completely, they are described as barrier reefs. Thus a ranking of reef structures from youngest to most mature would be: coral colonies, micro-atolls, patch reefs, fringing reefs, barrier reefs, and atolls.

In areas of optimum reef development, barrier reefs usually develop well offshore, and are often important in protecting islands from coastal erosion by waves. Two of the largest barrier reefs are the Great Barrier Reef off the northeast coast of Australia, more than 1,930 kilometers (1,250 miles) long, and the huge barrier reefs off the eastern coasts of Honduras and Nicaragua in Central America. There are 261 atolls in the world oceans, a large majority of which are in the tropical Pacific.

CORAL GROWTH AND EVOLUTION

To appreciate fully the constraints under which corals live, it is necessary to understand that they depend upon symbiotic algae called zooxanthellae that live within their tissues. Zooxanthellae are single-celled dinoflagellates that aid corals by providing nutrients and absorbing carbon dioxide. Both these processes speed the growth of corals by increasing the rate at which limestone is secreted. Corals without zooxanthellae grow only about 10 percent as fast as corals that contain the algae. Because the zooxanthellae live only in the well-lit upper levels of the sea, reef-building corals (hermatypic corals) also occupy only these shallower

CROSS-SECTION OF A CORAL ATOLL

levels—the coral euphotic zone. Many other kinds of coral, of course, exist at greater depths, some of the best known being the precious red, pink, black, and gold corals that are used to manufacture coral jewelry. Corals that live in the deep sea all lack zooaxanthellae, and are collectively referred to as ahermatypic corals.

While the evolutionary history of coral reefs and islands stretches back over 400 million years, when tabulate and rugose corals evolved during the Paleozoic era, most living reefs are only 5,000 to 8,000 years old. This seeming paradox can be explained by the fact that until 8,000 years ago, the sea level was over 30 meters (100 feet) below its present level. About 18,000 years ago, sea level was even lower, between 120 and 140 meters (390 and 460 feet) shallower than at present. This comparatively low sea level was the result of the last ice age, which caused the oceans to shrink in response to global cooling and the build-up of ice at high latitudes and elevations. Hence, all today's living reefs have formed only during the past several thousand years as rising seas flooded the shelves of elevated landforms. Many of their foundations, of course, were built of older fossil coral limestones which now underlie the latest episode of reef growth. In most parts of the world this modern veneer is less than 30 meters (100 feet) thick.

The history of rising sea level over the past 18,000 years also explains why many reefs are virtually drowned: that is, they are below the depth (approximately 30 meters/100 feet) where they can successfully regrow to the surface. Most coral reefs grow upward at less than 1 centimeter (½ inch) per year, whereas sea-level rise resulting from glacial meltwater and warming over this period has frequently exceeded this growth rate.

Ecosystems under Threat

Coral islands and reefs harbor some of the most complex and diverse ecosystems in the world. There are between 500 and 1,000 species of tropical reef fishes, and if all number and kind of algae, invertebrates, and microbes on the reef were to be fully enumerated there could be over 50,000 in any one major zoogeographic province.

Sometimes bizarre in appearance but almost always beautiful, reef ecosystems are often viewed as "delicate" or in "delicate balance" with nature. In truth, this is

Jeannie Mackinnon/Planet Earth Pictures

The beauty of coral reefs draws large numbers of tourists to coral islands. While the tourist industry boosts island economies, reef ecosystems often suffer as a consequence.

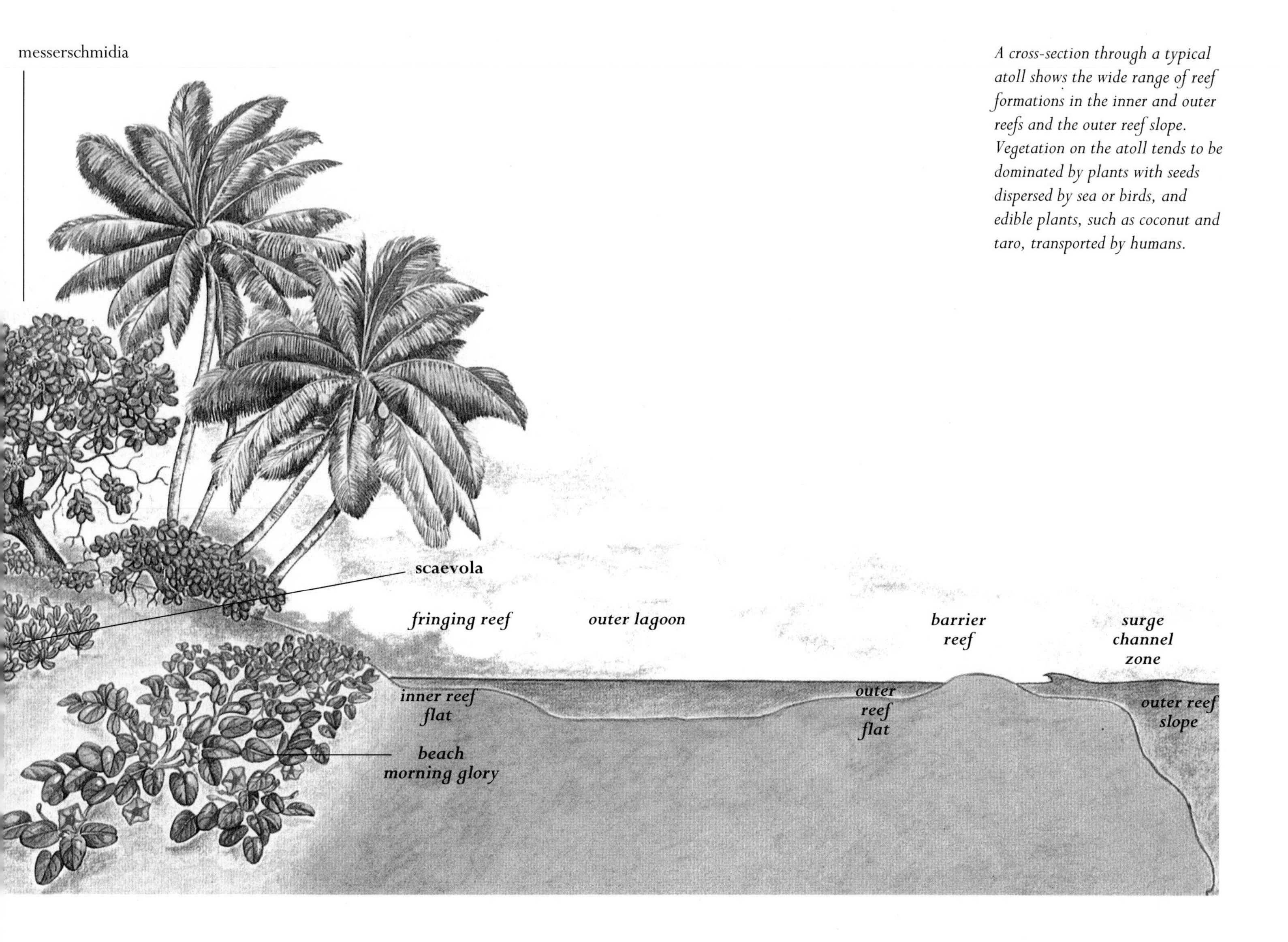

A cross-section through a typical atoll shows the wide range of reef formations in the inner and outer reefs and the outer reef slope. Vegetation on the atoll tends to be dominated by plants with seeds dispersed by sea or birds, and edible plants, such as coconut and taro, transported by humans.

LIVING TOGETHER

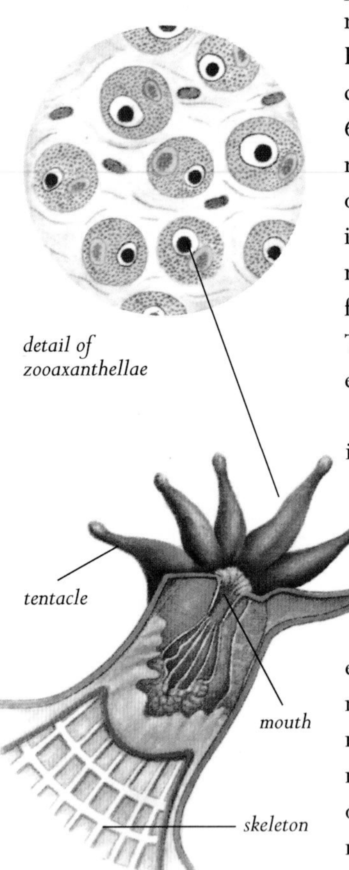

A cross-section of a coral polyp. Reef-building corals live in symbiotic associations with single-celled algae called zooaxanthellae. The coral provides a home for the algae, and the algae manufacture food by photosynthesis, some of which is shared with the coral.

Opposite. *An assortment of soft corals* Dendronephthya *on the Rainbow Reef, Fiji. This section of the reef appears relatively unspoilt, but many of the shallow corals on the reef have been devastated by cyclones, earthquakes, and infestations of crown-of-thorns starfish.*

probably nothing more than a well-accepted myth. Coral reefs have suffered and survived through eons of major Earth change, including episodes of mountain building, continental drift, and massive extinctions like the one 65 million years ago that wiped out the dinosaurs. More recently, they have survived the ice ages and the ravages of tropical storms. In fact, most of today's coral reefs are in one stage or another of recovery from some recent major disturbance, among them storms, sedimentation from river runoff, predation from starfish, or disease. The coral islands that depend on reef ecosystems are equally vulnerable to disturbance.

In spite of all this, coral reefs have managed to evolve into a spectacular assembly of specialized life-forms. While we now know that corals are quite hardy in evolutionary terms, it is nevertheless important to recognize the limits of coral reef ecosystems. These are limits associated with humankind: pollution, overfishing, habitat destruction, and exploitation. Coral reefs are slow growing and slow to recover from both natural and human-induced change. A reef may take between several decades and a century to recover fully from a devastating hurricane or typhoon, or the practice of using dynamite to harvest fish from the reef, which is widely used in the Philippines.

In many Third World countries in the tropics, coral-reef ecosystems have been under siege for several decades. Resources are being depleted at an alarming rate. Overpopulation, combined with poverty and the inability of government and society to manage people dependent on reef resources at subsistence level, pose a major threat to coral reefs on a global scale. Here is another example of the "Tragedy of the Commons" where technology may not provide a solution.

The threat to coral reefs is part of a larger problem of humankind versus nature. The question of which will prevail holds the answer not only to the future of coral reefs but perhaps even to the future of our own species.

As we approach the end of the twentieth century, and indeed the end of a millennium that has seen the human species explode from tens of thousands to billions of people, we stand at a true environmental crossroad, not only for coral reefs but for many species of plant and animal life. What we decide in the 1990s could determine the future ecology of the next century. In 1987 the world population exceeded five billion people. By the year 2000 our population will be six billion. If this rate of increase continues, we will almost certainly outstrip the capacity of planet Earth to support and maintain an ecological balance. The decline of coral reefs worldwide caused by human urbanization may already be trying to warn us that for either of us to survive, humankind must discover a way to control population growth. ■

Gary Bell/Planet Earth Pictures

THE BIRTH AND DEATH OF THE HAWAIIAN CHAIN

Of the approximately 100 Hawaiian islands born in the last 70 million years, 13 of these are still above sea level. The others have long since drowned in the gradual ride northwestward on the Pacific Plate. After each island is severed from the Hawaiian hot spot, it begins to subside and erode. Subsidence is caused by cooling of the underlying plate. Erosion is caused by wind and rain. In about 10 million years, these processes reduce most Hawaiian islands to broad foundations near sea level. It is on their shallow-water flanks that coral reefs take hold.

Krafft/AUSCAPE

Lava spurts and flows from Hawaii's Mauna Loa during an eruption in March 1984. Basaltic volcanoes, such as Mauna Loa, typically produce highly mobile lava that moves at up to 50 kilometers (30 miles) per hour.

Eventually, as the volcanic edifaces continue to subside, the upward growth of coral maintains the islands at sea level. In Hawaii, five such islands are in this stage of their evolution: Laysan, Lisianski, Pearl and Hermes Reef, Midway, and Kure islands. These islands are either atolls (barrier reefs surrounding lagoons) or simply coral islands in which the lagoons have filled in with limestone debris. Were it not for coral reefs, all of them would have drowned millions of years ago, and the Hawaiian Islands would be about 500 miles shorter, ending at Gardner Pinnacles. Today Gardner Pinnacles marks the northernmost outpost of the volcanic chain—a mere 30 meter (90 foot) high remnant of a once-majestic island. However, because of the coral reefs that surround Gardner, the island has at least five million more years to live before it drowns. After that it will be carried north into cooler waters where its surrounding reefs will drown, and Gardner Island will disappear beneath the sea. The threshold latitude where coral growth ceases and where coral islands drown, is called the Darwin Point, and no Hawaiian islands exist farther to the north. Today, this point in the chain is occupied by Kure Atoll.

The chain of volcanoes continues underwater for 2,400 kilometers (1,500 miles) more to the northwest, one drowned coral island after another all the way to Kamchatka. Collectively, these drowned islands are known as the Emperor Seamounts. The oldest Hawaiian island (72 million years) in the

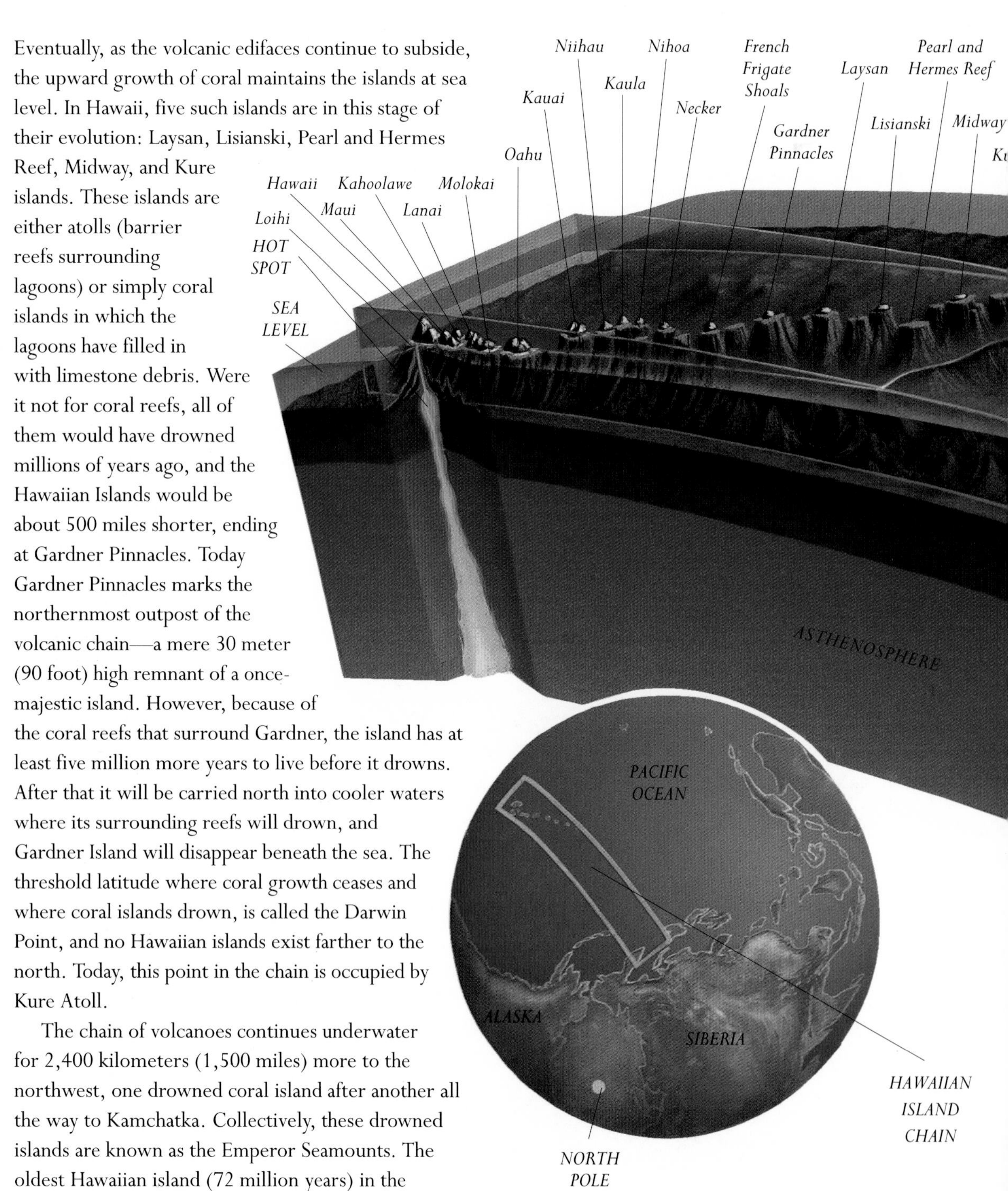

Emperors is called Meiji Seamount. It is the deepest and most northwest ediface, and is presently perched on the edge of the juncture between the Kuril and Aleutian trenches, undergoing subduction. Meiji would probably have been an atoll or a coral island for perhaps five million years before reaching the Darwin Point at 30°N. Since that time it has been an ever-sinking guyot and is now poised on the verge of ultimate destruction—subducting and remelting within the mantle of the Earth.

And so it is, from birth to death for every Hawaiian island and for every Hawaiian coral reef. The reefs live only so long as their foundations support them at or near sea level. Once below about 100 meters (330 feet), their fate is sealed and they become fossil limestones. Like crumbled cities of an ancient civilization, their remains tell of what was once a glorious time in which life flourished.

But the cycle continues and the newest Hawaiian volcano, Loihi, still submerged beneath the sea, will form a new island, with perhaps 30 million years to live before drowning at the Darwin Point. And in 70 million years it too will subduct and return to that from which it came. •

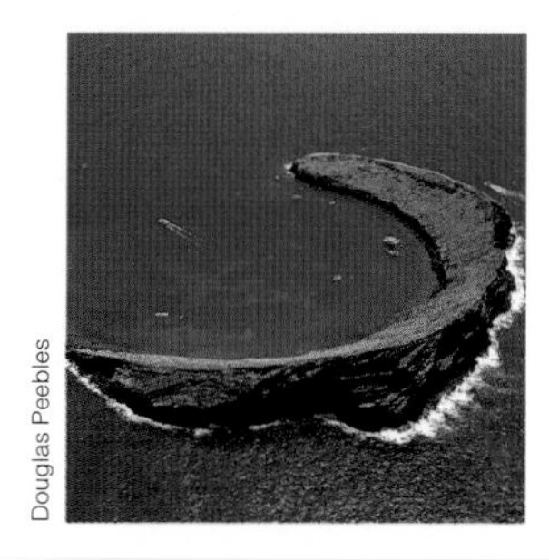

The fringing coral reefs of tiny Molokini, near Maui, will maintain the sinking island at sea level for a few million years to come.

The Pacific Plate moves northwestward at a rate of 10 centimeters (4 inches) per year. The older volcanic islands slowly subside into the sea and after about 70 million years will be subducted.

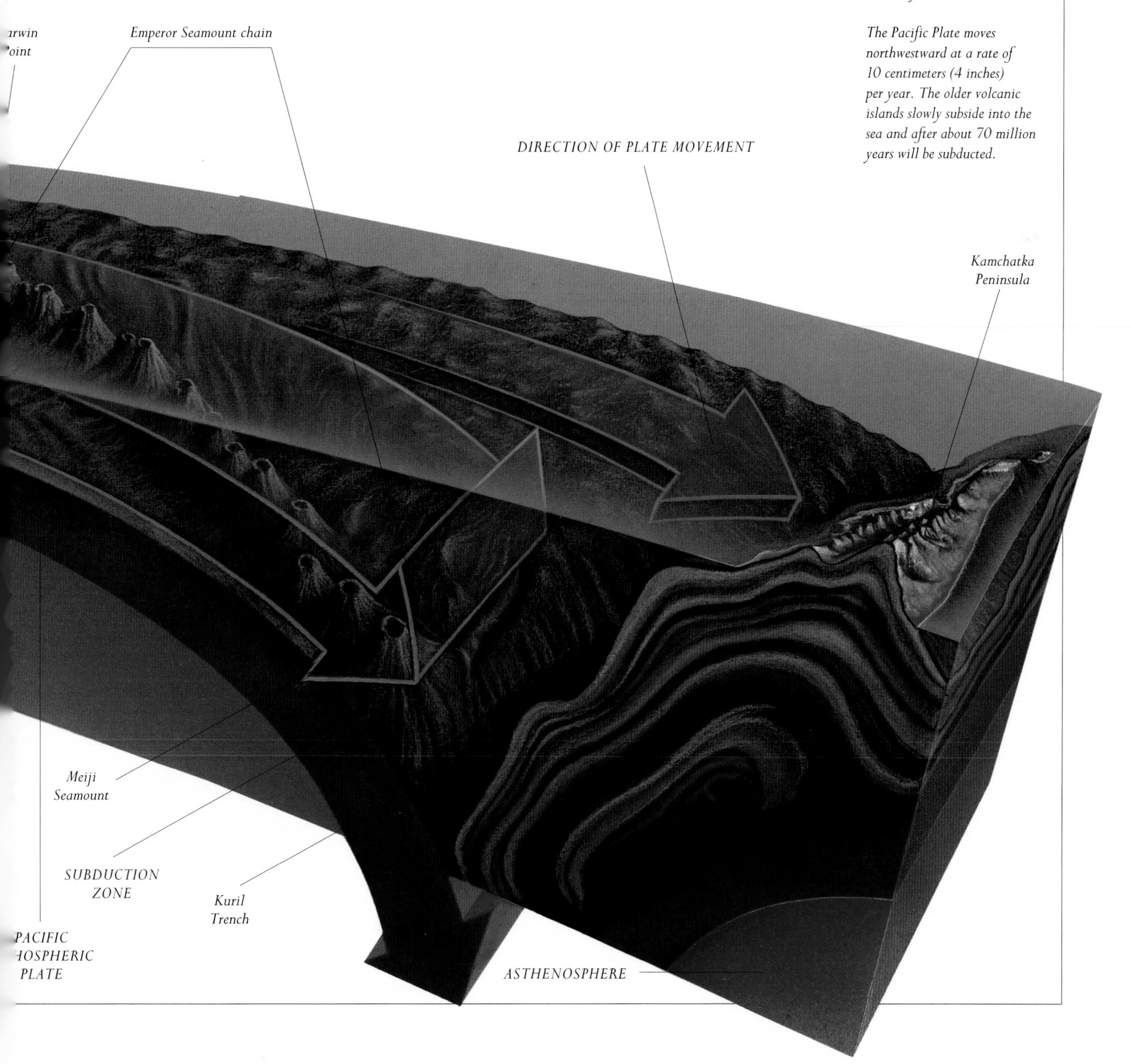

Tui de Roy/ AUSCAPE

PART TWO

Laboratories of Life

The conclusions Charles Darwin drew from his visit to the Galapagos Islands in 1835 changed scientific thinking about evolution. Islands are natural laboratories in which we can study the processes that contribute to the adaptation and diversification of life.

5 Colonization of Coral Islands

HAROLD HEATWOLE

Coral islands are not always what they are popularly thought to be. Some are indeed paradises, complete with swaying palms and golden beaches. Others are lonely, spray-swept tracts of barren sand, featureless and shifting. On the Great Barrier Reef of Australia both ends of this spectrum, as well as nearly all conditions between, can be found. Why are islands of the same region so different? Part of the answer lies in an understanding of two influences: time and stability.

Time and Tide

Continental islands were once part of the mainland, and their fauna and flora consist partly of species left behind when the sea rose and isolated them. By contrast, coral islands receive all their species by dispersal over water. Some species cross expanses of sea easily; others do so only with difficulty. A newly formed island is colonized quickly by easily dispersed species, and in time accumulates other species that find traveling harder.

The opportunity to accumulate species is not the only influence of time. Islands alter with age, and conditions become suitable for species that could not have survived had they arrived in an earlier period. However, improved conditions for some species are unfavorable for others. If an island is stable, eventually a balance is struck between the establishment of new species and the disappearance of old ones. If it is unstable, numbers of species may either increase or decrease depending on whether the island's environment is improving or deteriorating. The best way to understand these processes is to follow the fortunes of an island from its first appearance above the waters of the reef, to see how it becomes clothed in vegetation and populated with animals, and to witness the dramas played on its shores.

Peter Parks/Oxford Scientific Films

One of the earliest colonists of newly formed cays, and widespread and numerous on many Pacific islands, is the giant centipede Scolopendra, *which preys on insects and small beach scavengers.*

Opposite. *Vulnerable to cyclones and freak high tides, small coral sand cays may go through cycles of colonization and destruction, complete or partial, as in this tiny cay off Bora Bora, Tahiti, which has been denuded of almost everything but coconut palms.*

A Barren Platform

Sandy coral islands may form gradually, as minute particles of coral, broken shells, or the lime skeletons of algae and protozoans accumulate in the quiet waters of lagoons or in the lee of reefs. These sediments build up toward the surface until exposed at low tides. Then the wind heaps them into dunes tall enough that even high tides do not cover them. A submerged sandbar thus becomes an island. Other coral islands, called shingle cays, have a more tempestuous origin. Fierce tropical storms wrench blocks of living coral from the reef-face and hurl them into heaps of coarse rubble in the shallow reef-flat.

Regardless of the texture of an island at its inception, it is devoid of terrestrial life—it is a dead platform above the sea waiting to become vitalized. There is no shade, and the sands are hot, parched, and often laden with a burden of salt from sea spray, or from storm waves crashing over the beach. There are few soil nutrients and little organic matter. Rain quickly evaporates or is diluted by sea water that soaks the ground. Sands, unbound by vegetation, are at the whim of the wind, being shifted to and fro or blown to sea. It is this hostile environment that awaits the first terrestrial life cast upon an island's shores.

It is not surprising that most species reaching an island in its infancy do not survive there. Only the hardy few colonize and reproduce. Aptly they are called "pioneer species".

The Pioneers

A knowledgeable person could make a sensible prediction of the order in which species appear on an island. Green plants are the basis of all other life, so one might suppose them to be the first successful invaders. Without them other species could not survive. Next to arrive surely must be herbivores, animals that eat green plants, and then predators and parasites that consume other animals. Somewhere along the line, decomposers and scavengers living on dead bodies or wastes should appear.

As reasonable as these guesses are, they are not the whole story. In fact, animals often become established

Jean-Paul Ferrero/AUSCAPE

Philip Rosenberg/AUSCAPE

Lavena Beach, Taveuni, Fiji. Typical coral beaches such as this have a zone of ground-hugging plants such as Ipomoea *just above the beach, with fringing coconut palms and more established woodland behind.*

Above center. *Morning glories of the genus* Ipomoea, *like this one growing in Hawaii Volcano National Park, number about 500 species, of which several are pioneer colonists of tropical islands.*

earlier than plants. Many tiny coral islands on the Great Barrier Reef completely lack vegetation, yet have up to a dozen species of terrestrial animals, including earwigs, slaters, flies, beetles, spiders, and centipedes. These islands are not defying basic ecological laws. Their animals merely import food from the marine environment, and ultimately depend on the plants of that habitat for their sustenance. Food of marine origin arrives by several means. An important one is the wash-up of carrion. All beaches have their complement of dead fish, crabs, algae, jellyfish, and starfish. A feast awaits a terrestrial scavenger that might happen to arrive. Indeed, scavengers usually are the first terrestrial animals to colonize an island. Common ones on the Great Barrier Reef are the seaside earwig *Anisolabris maritima* and a small white-winged gnat *Loptocera fittkani*. The earwig burrows under carrion on the beach and eats it from below. The gnat hovers over it, landing to feed and lay its eggs on the decaying flesh.

Seabirds, too, are an important means of transferring food from sea to island. They eat fish or oceanic invertebrates, and use islands as platforms for roosting and nesting. Scraps of food brought to nestlings, bird guano, and the dead bodies of birds are materials ultimately of marine origin, but become available to island life via seabirds. For this reason, seabirds are called "transfer organisms".

Fungi and bacteria decompose dead organisms and break them down into finer particles. Terrestrial

Coconuts, Seychelles. Like a number of other plants, the coconut palm has seeds that float, reaching new islands by drifting on ocean currents.

animals, such as slaters and some insects, feed on this decaying organic matter or on the fungi that grow on it. These, too, make an early appearance on coral islands. As scavengers become abundant, the stage is set for predators to enter the scene. Large wolf spiders *Lycosa* species and centipedes *Scolopendra* species reach coral islands early and often become part of the food web before the arrival of green plants.

Although green plants usually arrive later than these animals, many are truly pioneers. These species are highly tolerant of salt, are drought-resistant, and can grow where there is very little nutrient in the soil. In addition, they cope with a continually shifting substrate. Wind can blow sand away from a plant and uproot it, or it can cover and smother it. Pioneer plants have two strategies for combating these hazards. One is to grow rapidly and produce many seeds as very young plants. The future of the species on the islands is then assured by the abundant seed that has been scattered. A good example of such a plant is swine-cress *Coronopus integrifolius*. During winter, when rain is most abundant, it sprouts from seed left from a previous year, flowers, sets fruit rapidly, and then dies.

The other way of coping with unsuitable substrates is to send out runners or vines. These spread over the sand and put down roots. Subsequently, if part of the plant is either excavated or covered by sand it will not die, as unaffected parts can continue to supply water and food. Common examples of such plants on coral islands of the Great Barrier Reef are the beach

Wolf spiders are common on many remote Pacific islands. The female of the family Lycosidae carry their young around on their backs until the young are old enough to fend for themselves.

A newly formed island is colonized quickly by plants and animals that are dispersed easily by air or water. Some are carried by wind, some hitch rides in the feathers of birds, some drift on the sea and some ride on drifting logs and other debris. Seabirds carry many kinds of organisms tangled in their feathers or in mud caught in their feet.

morning glory *Ipomoea pescaprae brasiliensis*, the beach pea *Canavalis maritima*, and the bird's beak grass *Thuarea involuta*.

Animals disperse to islands by air and by water. Some, such as insects, birds, and bats, can reach islands under their own power. But even within these groups many species are weak fliers and would rarely fly the long distances required to colonize remote islands. These, and even some small flightless animals like mites, may be blown passively by the wind after being wafted aloft on rising air currents. The young of some spiders are particularly adept at this: they spin small parachutes of gossamer that catch the wind and carry them away. Large centipedes and groundspiders travel by sea, usually rafting in or on a floating log, coconut, or other land debris. Even snakes and lizards have been seen at sea riding on floating objects.

Plants, too, are transported by sea currents, and beaches are often littered with seeds and fruits. Pioneer species are resistant to sea water and survive prolonged immersion. On reaching land after a long sea journey, they can still germinate. With the establishment of green plants, arriving herbivores have a good base and can become established on the island.

The tiny island is no longer bare. It has a small community of arthropods, based mainly on a scavenging industry, and a sparse covering of pioneer plants with a few herbivores.

LATER SETTLERS

A small island is essentially all beach. However, as more sediments accumulate, wind-driven sand is piled higher, and the beach expands outward. Stormy seas no longer wash over the whole of the interior, and pioneer vegetation begins to stabilize the soil and reduce wind erosion. The greater accumulation of sand retains rain, which floats on top of the salty water in the soil. Plants that are less drought-resistant and less salt-tolerant than the pioneer species can now colonize. Seabirds prepare the soil by fertilizing it with guano.

Almost all the next-stage settlers are dispersed by birds. Sometimes their seeds or fruits are attached to feathers; sometimes they are eaten by a bird, carried in its digestive tract, and deposited in excrement on a distant shore. Many of these plants have adaptations that facilitate their transport by birds. The chaff flower *Achyranthes aspera*, for example, has fruits with sharp hooks that cling to feathers, and the fruits of the tar-vine *Boerhavia diffusa* are covered with a sticky gum that enables them to adhere.

The beaches are still buffeted by waves, swept by wind, and sprayed with salt water, and only pioneer species grow there. But farther inland, the changes wrought by seabirds and greater isolation from the marine environment produce a milder, more stable habitat. The island becomes zoned into two habitats: a peripheral one of pioneer vegetation and hardy

HITCHING A RIDE

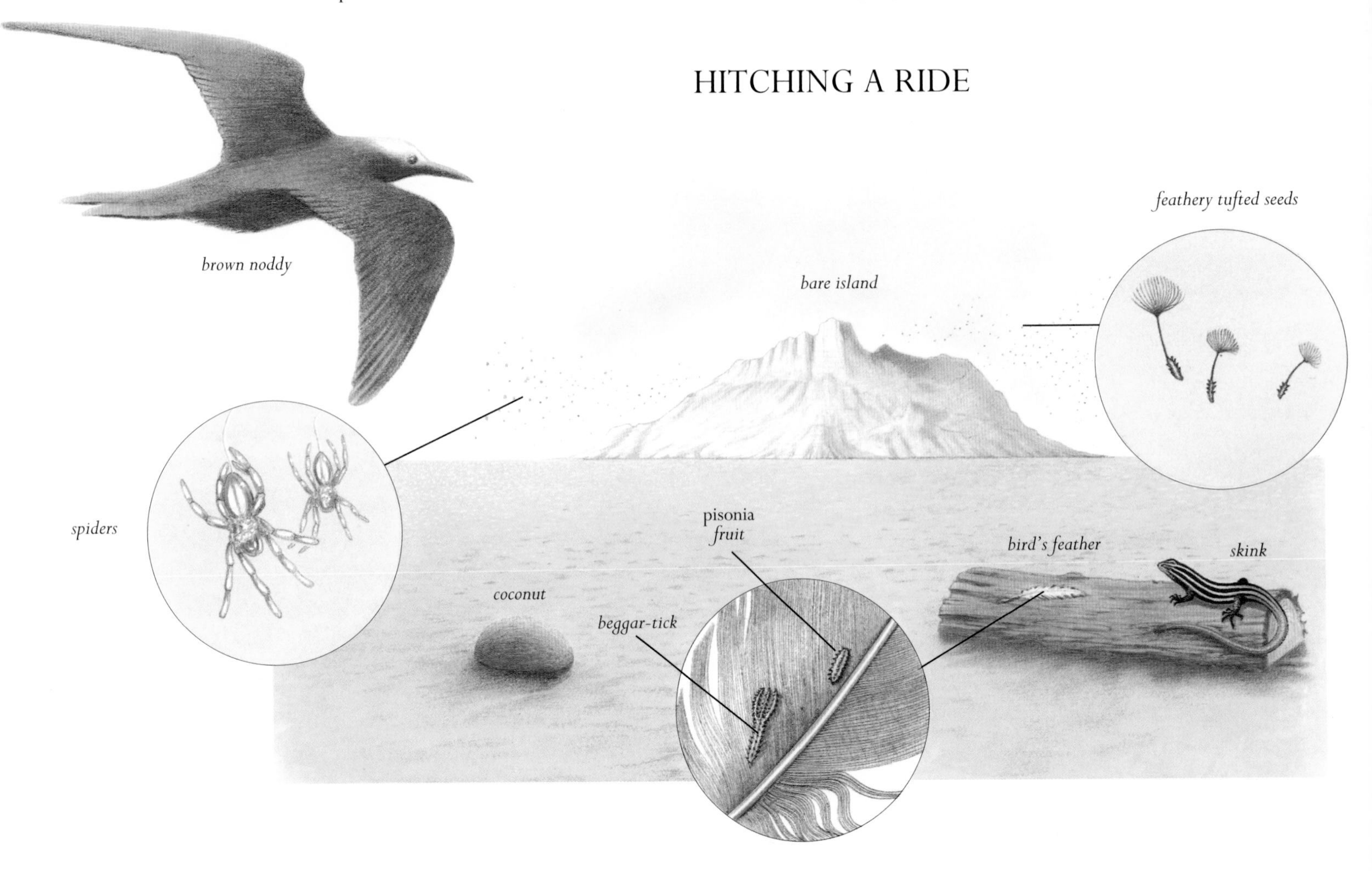

animals, and a central one of less hardy species. In the central zone, vegetation grows more densely and luxuriantly. Its shade cools the ground and reduces loss of moisture by evaporation. Soil is sheltered from the wind and is stabilized through binding by roots. Dead leaves and stems provide a covering of moist litter. More species of plants now find a favorable seedbed, grow, and support a larger number of herbivores.

The island's central zone is now a place where herbivores play a greater role in the food web.

ERECTING THE BARRICADES

So far the island supports only low vegetation such as herbs, grasses, and vines. Further changes depend on the arrival and establishment of shrubs. The most important shrub on the islands of the Great Barrier Reef is the sea-dispersed octopus bush *Argusia argentea*. Its seeds require soaking in salt water before they can sprout. After a period of exposure to salt, the seeds germinate when soaked with rain, often on a distant beach to which they have been carried by ocean currents for hundreds, or even thousands, of kilometers. The sea lettuce tree *Scaevola sericea* is another plant that becomes established at this time.

At first there are only a few bushes scattered around the edge of the beach. Gradually, the gaps close until a ring of shrubs surrounds the island. Following the shrubs come arthropods that graze on their leaves and imbibe nectar from their flowers. Flies and moths become more abundant, and predators that feed on them become established.

The island now has three zones: a peripheral one of sparse, pioneer vegetation, and a shrub ring, inside which is a central zone of dense herbiage.

INVASION OF THE TREES

A shrub ring has a profound effect on the interior environment of an island. It serves as a windbreak and screens out salt spray, allowing more species of plants, including additional shrubs and even some trees, to become established in this sheltered habitat. These are accompanied by an ever-increasing number of insects and other animals. As occasional wandering land birds are attracted, species carried as seeds in their digestive tracts become established. Once such plants are numerous enough to provide sufficient food, small populations of land birds become resident. In time, an open parkland of low vegetation interspersed with scattered trees and shrubs replaces the previous central zone of low vegetation. This provides additional shade and a cooler, moister environment. Gradually, as more trees grow, parkland gives way to dense forest.

On the Great Barrier Reef a common forest tree is the pisonia *Pisonia grandis*. It has a special relationship with seabirds, especially the black noddy *Anous minutus*. The large amount of fertilizer required by pisonia trees is supplied by guano from the noddies nesting on their branches. Dense shade cast by the trees and high levels of guano in the soil prevent the growth of most plants, and the low vegetation is lost. A thick layer of decaying leaf litter accumulates and a richly organic soil develops. This teems with numerous life-forms including mites, slaters, centipedes, and springtails. With less open space, there are fewer ground-nesting terns, but muttonbirds such as *Puffinus pacificus* come instead to dig their nesting burrows in the soil.

The island is now very different from the original desolate sands that first appeared above the sea. It has a number of vegetation zones, each with its characteristic fauna. There is still a peripheral beach of pioneer species, bordered by a shrub ring. Inside the ring, part of the island may be open parkland and part forest. Sometimes forest completely replaces parkland. Ground-nesting seabirds have given way to burrowing muttonbirds, and to tree-nesting noddies and land birds. We have reached our paradise. Can such verdure be sustained?

Jean-Paul Ferrero/AUSCAPE

Unlike most of its relatives, the black noddy Anous minutus *nests in trees. Mostly it selects* Pisonia, *a plant colonist whose sticky seeds tangle in the birds' plumage, to be carried from island to island.*

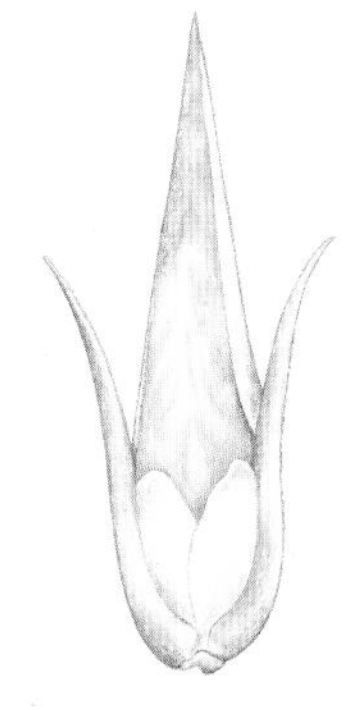

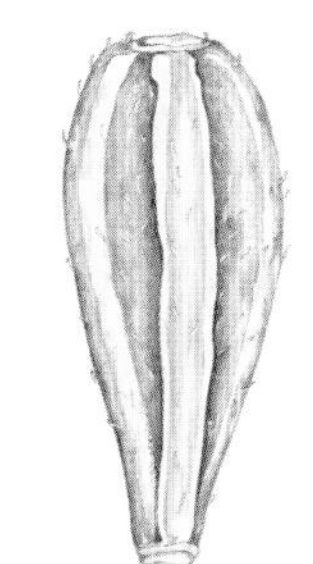

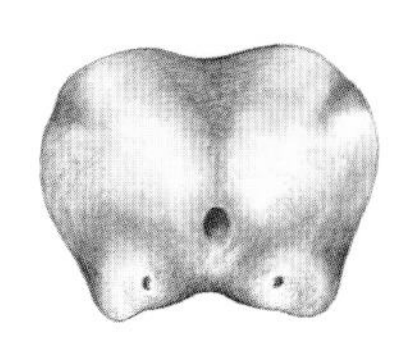

Three examples of plant dispersal strategies: seeds of the chaff flower Achyranthes aspera *(top) have sharp hooks; seeds of the tar-vine* Boerhavia diffusa *(center) are coated with sticky gums to attach to bird feathers; and seeds of the octopus bush* Argusia argentea *(bottom) must soak in sea water before they can germinate.*

Paradise Lost

So far the discussion has centered on one of the two major ecological influences mentioned at the beginning: time. Now it is appropriate to consider the second factor: stability. The stages of island development are not irreversible. At any point islands can deteriorate to previous stages, or even disappear altogether. Setting off such changes are storms, erosion, seabirds, sea turtles, and the activities of humans.

During severe tropical storms, violent seas pound over beaches, flooding vegetation and animals with sea water, or covering them with sand. Soils on coral islands provide a record of these events. They are often layered with dark bands of organic soil alternating with light bands of mineral sand. Each dark band represents an old surface that was once vegetated, and

Kathie Atkinson\Oxford Scientific Films

Plant pioneers must cope with extremes of sun, wind, sand, and salt. One of the hardiest of these is the bird's beak grass Thuarea involuta, *common and widespread on beaches along the Great Barrier Reef, Australia.*

each light one a smothering by sand. This repeating vegetation pattern tells of many cycles of destruction and re-establishment.

Even in the absence of storms, beach sand is washed from place to place, and small alterations of the coastline occur, especially at the narrower ends of an island. A time-lapse film would show the tip whipping back and forth like a tail. It is this ceaseless activity that keeps the edges of an island at the pioneer stage of colonization. Stable islands stay in the same place; erosion and deposition merely affect the edges. However, when there are changes in the currents around an island, the erosion–deposition cycle no longer nibbles around the edges, but whole portions are eaten away. Rapid erosion may cut one side of an island back into what had been its central zone. At the same time, on the opposite side of the island, redeposited sand forms new, expanding beaches open to colonization of pioneer species. The zones become skewed, with the central zone offset toward the eroding edge. When one side continues to erode and the other to expand, an island "walks" across its reef. Sand may be washed out to sea and the island dwindle in size, eventually being left only with pioneer vegetation. An island near the rim of its reef may go over the edge and disappear altogether.

Seabirds not only play an important role in the development of an island, they can also contribute to its degradation. Where birds nest in dense colonies, their trampling and high concentrations of guano kill plants. Bare patches develop and expose the sand to wind. Often colonies of birds move from place to place on an island, or from island to island, in subsequent breeding seasons. Such "rotation" allows the recovery of vegetation. However, if populations are too large, an entire island may become bare, and any plant hardy enough to germinate is destroyed as soon as it appears. Gannet Cay in the Swain Reefs of the Great Barrier Reef is such an island. In the late 1960s it was large and had a lush cover of plants. Thereafter, it began moving over its reef and became progressively smaller. Its dense breeding populations of brown boobies *Sula leucogaster* and masked gannets *Sula dactylatra* have destroyed the vegetation, and now the cay is completely bare.

A loggerhead turtle Caretta caretta. *Although they visit only when breeding, marine turtles play their part in the dynamics of a coral island, disrupting vegetation as they come ashore to dig their nests and lay their eggs in the sand.*

Len Zell/Oxford Scientific Films

Sea turtles also destroy vegetation. Females dig pits in the sand in which they lay their eggs. Plants growing at the nest site are uprooted and may be killed. Some pioneer species, such as bird's beak grass, beach morning glory, and beach pea, survive better than others, being vines or creepers adapted to shifting substrates. In this case turtles, rather than wind, are causing instability, but the result is the same—plants unable to cope with moving sand do not survive. Therefore, turtle nesting areas in central vegetation revert to a pioneer stage.

Humans can have a great effect on coral islands. Guano miners removed plants and topsoil from islands of the Great Barrier Reef at about the turn of the century. This was followed by the introduction of goats, which in turn took their toll of the vegetation. Although goats are no longer present and guano mining has ceased, some islands still suffer from the consequences of these disruptions. With the development of tourism, there has been further clearing of vegetation, construction of buildings, and the introduction of exotic plants. Weeds and insects are introduced inadvertently, especially in sand or gravel brought in for construction purposes. Waste disposal can cause acute environmental problems.

Humans may have an indirect, as well as direct, ecological effect. As well as transporting weed seeds directly in their gear or attached to their clothes, people may aid the introduction of weeds by attracting gulls. Gulls eat a variety of foods, including weed seeds, and thus transport weeds to new localities. They are also scavengers and congregate in the vicinity of humans. Up to three-quarters of the gull population of the entire Bunker–Capricorn group of islands has been drawn at one time to dumps on Heron Island. With a resort, a marine research station, and a park headquarters, this island has a high human population. The number of weed species has progressively increased over the years. It is still uncertain how much of this is due to seeds being carried by people directly, and how much to dispersal by gulls attracted to the human settlement.

Some undesirable changes have occurred through lack of understanding of the ecology of islands. Many, however, have resulted from greed, coupled with a callous disregard for the long-term reef environment. Developers have been quick to exploit before adequate

THE DEVELOPMENT AND DEGRADATION OF CORAL ISLANDS

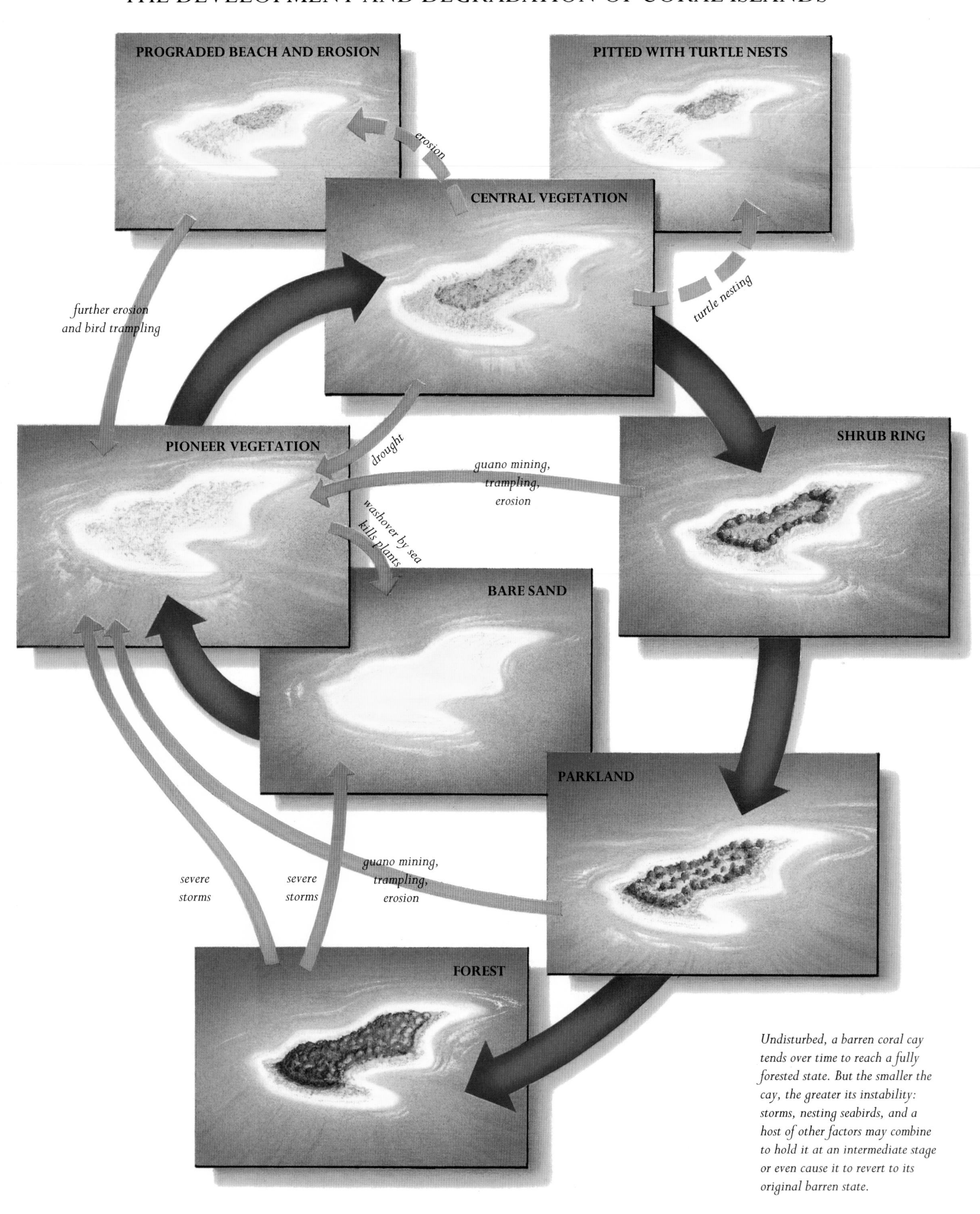

Undisturbed, a barren coral cay tends over time to reach a fully forested state. But the smaller the cay, the greater its instability: storms, nesting seabirds, and a host of other factors may combine to hold it at an intermediate stage or even cause it to revert to its original barren state.

A brown booby Sula leucogaster *with chick. Many seabirds nest on small coral islands in colonies, sometimes so large and dense that the birds may destroy all plant growth with their guano and nest-building activities.*

Opposite. *A coral reef exposed at low tide on Heron Island, Great Barrier Reef, Australia. Tourists thronging to explore such reefs may affect the delicate island ecology in many subtle ways, from the inadvertent introduction of weed seeds to encouraging scavenging gulls.*

Kevin Deacon/AUSCAPE

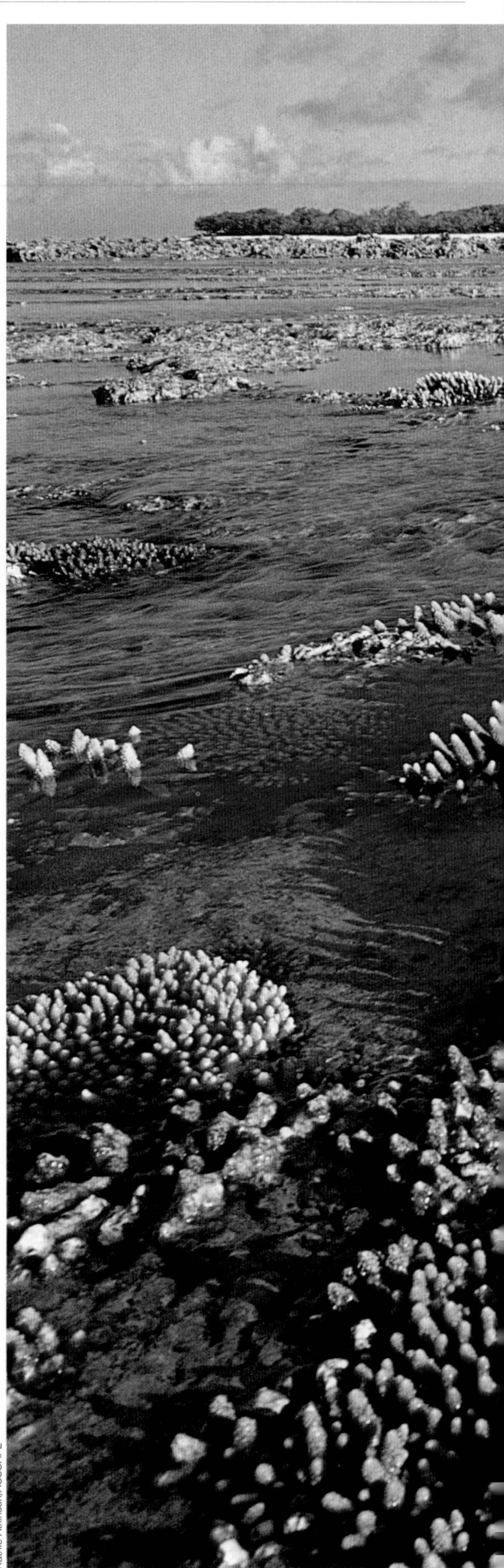

Kathie Atkinson/AUSCAPE

studies were carried out and safeguards established. In some cases government requirements have been ignored and restrictions exceeded.

Not all changes have been bad; there have been some success stories. Lady Elliot Island was perhaps the most ravaged island on the Great Barrier Reef, having been denuded of its vegetation and soil by guano miners and goats. Now it is one of the most pleasant of the islands. It has been carefully restored through an enlightened reforestation program. There are shady groves of coastal she-oaks *Casuarina equisitefolia*, and seabirds are returning to nest in ever-increasing numbers. This miracle was achieved, not by public support and governmental initiative, but by a private citizen wishing to establish a modest, comfortable holiday resort in natural surroundings.

There is a lesson in this. Human interference took a forested, stable island from the apex of its development back to a barren condition. It remained in that state for decades, until humans once again intervened, this time toward restoration rather than degradation. Islands are fragile and dynamic ecological systems. They develop from barren platforms to forested paradises, but they can also go in the reverse direction, and often do. Humans can be a force for improvement or an ecological catastrophe. So far, the track record has been poor, with human activities all too often accelerating destructive forces. By natural processes, islands come and go, develop and deteriorate. Humans have the power to tip this balance. Overexploitation could lead to deterioration of even the more stable, forested islands, leaving nothing to replace them—truly a paradise lost. ■

6 The Pacific Islands

E. ALISON KAY AND G.M. WELLINGTON

When the explorer James Cook rounded Cape Horn on his ship Endeavour *and ventured into the Pacific in February 1769, a new world of animals and plants was introduced to Europeans. Among the novelties they saw were tree-living land snails, birds such as the honeycreeper and toothed pigeon, and cultivated plants such as the breadfruit. In the course of the eighteenth and nineteenth centuries, shells, bird skins, preserved fish, and pressed plants were collected and taken to the herbaria, museums, and curiosity cabinets of Europe. There they opened new vistas to scientific knowledge, and continue today to give impetus to studies in the problems of distribution and evolution of plants and animals of oceanic islands.*

Rules of Island Life

The explorer-naturalists who traversed the Pacific two centuries ago were keen observers, recognizing not only new species but some of the peculiarities of island life. Johann Reinhold Forster, on Cook's second voyage, wrote that "The countries of the South Sea . . . contain a considerable variety of animals, though they are confined to a few classes only . . .", and he noted as an example that the only mammals he saw were "the vampyre and the common rat". Adelbert Chamisso,

First discovered in 1976, the Fijian fruitbat Pteralopax acrodonta *is so far known only from the small islands of Taveuni. Other members of the genus occur on the Solomon Islands but nowhere else—an intriguingly limited distribution for such a highly mobile group of animals.*

Opposite. *The spectacular Na Pali coast, Hawaii, is notable for its deeply dissected gorges, which are the product of very high rainfall falling on very steep slopes. These cliffs have since been mauled by an unusually savage hurricane that inflicted severe damage on the vegetation.*

the German poet and naturalist on the Russian ship *Rurik* in 1818, added his observations: "This rich Flora seems to have become more scanty in the islands of the Great Ocean, from the west towards the east", and ". . . the appearance of nature in the eastern islands of the South Seas, reminds us at once of Southern Asia and New Holland, and is wholly dissimilar to America."

The peculiarities remarked on by the explorer-naturalists can be restated as the rules of Pacific island life: the species that inhabit Pacific islands are few in numbers compared with those on equivalent continental areas; there is a gradual elimination of major groups of plants and animals from west to east across the Pacific; there are diminishing numbers of species from west to east; many islands have species that are endemic, or unique to them; and the animals and plants of Pacific islands are, for the most part, related to those of the west rather than the east.

The number of plant and animal groups falls off rapidly once the shores of the Philippines, Australia, and the Malay Archipelago are left behind. East of the Solomon Islands, which have a marsupial mammal and some rodents, there are only bats; and the 17 bat species in the Solomons are reduced to four in Fiji, and one in the Cook Islands. There are neither frogs nor snakes east of Fiji. Even the number of birds falls from west to east, from more than 250 land birds in New Guinea to nine in the Marquesas. Among the higher plants, conifers, bamboos, and rhododendrons do not penetrate beyond New Caledonia and Fiji. Mangroves and large fruited trees of shorelines, such as *Barringtonia*, reach the Marshall Islands but are not found in Hawaii.

The Western Fringe

Seven major island groups fringe the western border of the Pacific: New Zealand, New Caledonia, Fiji, Vanuatu, the Solomon Islands, New Guinea, and Palau. Beyond them to the west are the continents of Australia and Asia, the Malay Archipelago, and the Philippines, all far richer and more varied in plant and animal species. In the Malay Archipelago it is estimated there are 3,000 species of plants; in Vanuatu and New Caledonia there are fewer than 1,000 species. The islands of the western fringe are both continental

Three life-forms that exhibit three very different roles in the endless pageant of animal and plant dispersal: Fiji's crested iguana Brachylophus vitiensis *(below left) sometimes survives sheer flukes of transport, such as lengthy ocean voyages on floating rafts of leafy debris; the screw-pine* Pandanus tectonis *(below center) of French Polynesia, belongs to groups that routinely exploit wind and sea for maximum dispersal; and Goodfellow's tree kangaroo* Dendrolagus goodfellowi *(below right) exhibits almost zero dispersal.*

and volcanic; some of them, such as New Zealand, are very old; all are densely vegetated. Their plants and animals are the same as those of Australia, the Malay Archipelago, and Southeast Asia, with some elements unique to each island, and often with some major order or family missing.

Tropical rainforests are dominated by massive auracarias and kauris, broad-leaved breadfruits and durians, *Metrosideros*, palms, various species of *Pandanus*, and climbing vines such as *Freycinetia*. Shorelines are rimmed by mangroves. New Zealand, in a more temperate climate, has many tropical plants in its forests, including *Metrosideros* and *Pittosporum*, as well as non-tropical plants such as the southern beech *Nothofagus* and *Fuchsia*.

The western islands are a mosaic of presence and absence. New Zealand has three frogs but no snakes, yet both frogs and snakes are present in Fiji. There is a marsupial mammal in the Solomons but no native mammals in New Caledonia. New Zealand is legendary for its now-extinct moa, which at 3 meters (10 feet) high was the tallest bird in the world. In New Zealand, too, there is the lizard-like tuatara, and in Fiji an iguana. The kagu is endemic to New Caledonia. It is a remarkable grayish ground bird, about as big as a chicken, with a crested head, well-developed wings (although it never flies), and a call like the bark of a dog. In New Guinea there are tree kangaroos, cuscuses, giant spiny anteaters, birds of paradise, and cassowaries. The birds of paradise provide a perplexing example of absence: New Guinea is crowded with these distinctive birds and yet, remarkably, the neighboring Bismarck Islands have none.

Insects are big and colorful. Fiji is home to huge stick insects and one of the largest beetles in the world. Many of the beetles of New Guinea are a gaudy combination of gold and rose and green. The flight of the birdwing butterflies of New Guinea and the Solomons has been described as "poetry in motion". Some of the land snails are also spectacular. In New Zealand *Paryphanta*, a large mahogany brown snail, feeds on earthworms. In New Guinea *Papustyla pulcherrima*, the brilliant green tree snail of Manus Island, lives high in trees in the rainforest.

Jean-Paul Ferrero/AUSCAPE

Pete Atkinson/Planet Earth Pictures

HIGH ISLANDS OF THE PACIFIC

East of the western fringe are several volcanic islands among the array of islands recognized as Polynesia and Micronesia. Guam and others of the Mariana Islands, and Truk, Pohnpei and Kosrae in the Caroline Islands are north of the equator; the Samoas, Tongas, Tahiti, the Marquesas, and Easter Island are south of the equator. Because they rise steeply from the sea, often to heights of more than 300 meters (1,000 feet), these islands are called "high islands" to distinguish them from the "low" limestone islands among which they lie.

The woody plants, jungle climbers, mangroves, palms, *Pandanus*, and ferns of the forests are like those of the west, but here there are fewer species, even in the western islands such as Guam and Samoa; and the number of species rapidly declines eastward: Guam has about 356 species of native plants, Samoa around 320 species, and the Marquesas have fewer than 200 species. Among these islands a few genera are endemic: *Fitchia*, a relative of the sunflowers, is unique to Tahiti and nearby islands; on Bora Bora, one species of shrub grows only on the cloud-shrouded upper slopes; and Rarotonga has a unique species of large forest tree.

The dominant animals on these high islands are land birds, small insects, and mollusks; there are no mammals except fruit bats, and no amphibians or reptiles. Each island group has its own subspecies of fruit doves, fly-catchers, fantails, and reed warblers. In the Marquesas there are two species of pigeons and on Tahiti two species of kingfishers. The most remarkable of the fruit pigeons is the tooth-billed pigeon, a large bird endemic to only two islands in Western Samoa. In the Mariana Islands there is an endemic crow and on Guam a flightless rail.

A conspicuous feature of the insect fauna of these islands is the absence of large groups, families, and orders common to all continents: there is neither a single native species of scarab beetle nor any endemic leaf beetles. Few if any ants range east of Rotuma, Samoa, and Tonga. Anopheline mosquitoes are absent east of 170°E. The larger part of the insect fauna is made up of weevils. While there are comparatively few genera, some of them have disproportionately large

C.B. & D.W. Frith/Bruce Coleman Ltd.

The female birdwing Ornithoptera priamus *is among the largest and most spectacular of the world's butterflies.*

Jean-Paul Ferrero/AUSCAPE

Robert Harding Picture Library

One of the most extraordinary examples of adaptive radiation among birds is the family Drepanididae, which is confined entirely to the Hawaiian Islands. Nearly half of its approximately 22 member species have become extinct since the arrival of Europeans, and the status of most others is precarious. This, the crested honeycreeper Palmeria dolii, *still survives on Maui.*

numbers of species. In the Marquesas, for example, a relative of the leafhoppers is represented by eight endemic species.

Pacific island land snails are the only group in which entire families are endemic to the Pacific. The Endodontidae and Partulidae occur on all the high islands of the central Pacific, each island with its own suite of endemic species. There are three genera of partulids: *Partula*, distributed from Belau to the Society Islands; *Eua*, confined to Samoa and Tonga; and *Samoana*, found from the Marianas to the Marquesas. *Partula* reaches its greatest diversity in the Society Islands where there are 65 species, most of them apparently tree-dwelling.

HAWAII: A SPECIAL CASE

The Hawaiian Islands are the most isolated major island group in the world. With their low coralline islands in the north and major volcanic islands in the south, the landscape ranges from snow-covered mountain tops to shoreline desert, and from bog and rainforest to new lava flows. More than 90 percent of the flowering plants, insects, and land snails are endemic, but the biota is disharmonic, that is, it lacks major elements which occur elsewhere such as gymnosperms, bromeliads, figs, and mangroves; and there are only four orchid species, one genus of palms, and two genera of butterflies. There is one endemic family, the land snail Amastridae, and an endemic subfamily of birds. Among the unique flora are lobelias; tarweeds (which include Hawaii's most famous plant, the silversword); and legumes such as *Vicia*, *Erythryina*, and *Canavalia*.

Birds and insects have evolved in remarkable directions. The bills of the honeycreepers range from the delicate curved bills of nectar-sipping birds to the massive mandibles used in tearing off bark and wrenching open burrows of larval insects. There are some 500 endemic species of drosophilid flies, more than in any other part of the world. Tarweeds and beggars ticks, in different forms and different habitats, range from the top of Haleakala Crater on Maui to new lava flows on Hawaii.

LOW ISLANDS OF THE PACIFIC

There are more than 300 islands between Belau and the Tuamotus, most of which are low islands, either atolls or raised coral islands such as Makatea and Henderson Island. These islands rise only a few meters above sea level and are distinguished not only by their geological base of limestone, but by their sparse and cosmopolitan flora and fauna.

The beach plants of low islands are the same as those on the beaches of all islands throughout the Pacific—beach morning glory, *Pandanus*, *Scaevola*, and beach heliotrope, to name a few. On low islands, however, these plants represent the flora of the entire island. No two islands have exactly the same plants, and what plants there are may form a jungle where it is rainy and scrub where rainfall is low. The northernmost atoll in the Marshall Islands is in a semi-arid belt and has a flora of nine species; an atoll in the south, where it is extremely rainy, has perhaps 60 native species.

The birds of low islands are predominantly nesting seabirds which spend many weeks at sea. There is also a surprising number of land birds. Not only does each island have (or had) its own flightless rail, but fruit pigeons and reed warblers are also endemic to specific islands: the Wake rail, the Henderson Island fruit dove, and the Makatea fruit dove are examples.

THE SUCCESS OF CHANCE DISPERSAL

Most biogeographers now agree that plants and animals arrived on Pacific islands accidentally, the result of chance dispersal over water: rafting on floating trees, logs, and pumice; as planktonic larvae in ocean currents; as propagules (offshoots) uplifted by air currents; and as seeds and small animals stuck to the feet or feathers of birds.

There is convincing evidence of the effectiveness of chance dispersal. Darwin himself conducted several

William H. Amos

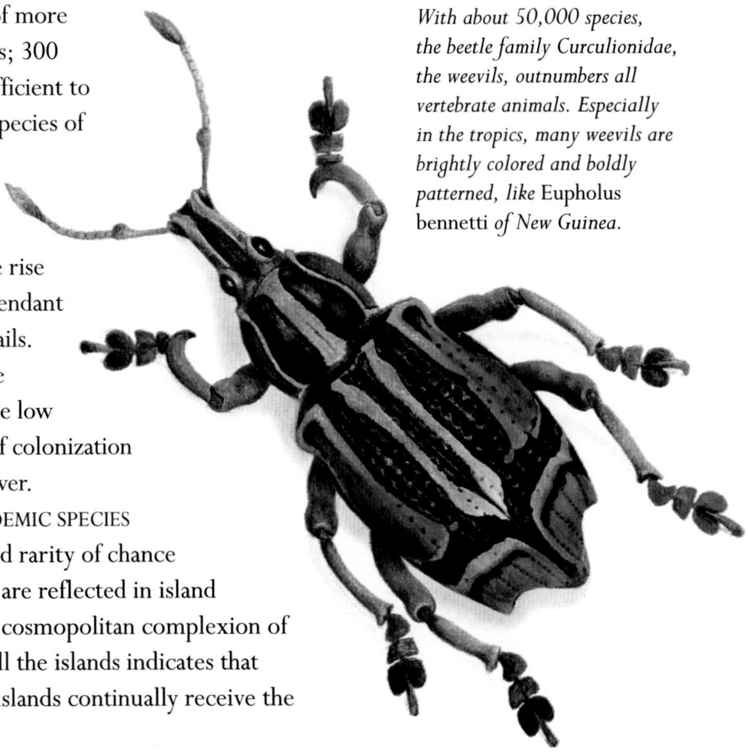

Some of Hawaii's landscapes consist of barren expanses of lava. Yet even in these areas, the process of colonization persists, here symbolized by a minute spider picking its way through a labyrinth of lava crystals.

experiments to see how long seeds would remain viable in sea water. A century later, J. Linsley Gressitt experimented by trapping insects in nets towed by planes flying over the Pacific, and Rudolph Scheltema sieved mid-Pacific water for larvae of marine animals. Gressitt found insects and spiders in the jet stream in the same proportions and of the same types as those on the islands below; Scheltema found larvae of coastal marine snails and worms hundreds of kilometers from land. The efficacy of natural rafts has been documented by Paul Jokiel who has found pumice, to which corals and other sessile (permanently attached) invertebrates cling, washed up on island shorelines throughout the Pacific.

Darwin argued that chance dispersal was like a filter, that only those organisms capable of dispersal would end up on distant islands. Thus oceanic islands are populated with a limited number of representatives of plants and animals, and the characteristics of Pacific island life derive, in part, from filtering out most continental species, and from the rigors of long-range dispersal. This explains the absence of all but such agile mammals as bats and seals, of freshwater fishes that cannot survive in the ocean, and of amphibians and those reptiles for which an ocean voyage is not feasible. And it explains what is found on Pacific islands: plants with seeds that can be dispersed by wind, or birds, or in the sea; birds that survive chance dispersal by storm or wind; and insects and snails that are easily transported by wind, or raft, or bird.

However they might have come, the indications are that in the right environment only a very few need have arrived. In Hawaii it is estimated that over a period of several million years only about 272 successful colonists would account for the indigenous flora of more than 1,000 species; 300 founders were sufficient to produce 10,000 species of insects; and 22 to 24 ancestral colonizations gave rise to the 1,000 descendant species of land snails. On less hospitable islands, such as the low islands, the rate of colonization may be much slower.

ENDEMIC SPECIES

The frequency and rarity of chance migratory events are reflected in island populations. The cosmopolitan complexion of beach plants on all the islands indicates that beaches and low islands continually receive the

With about 50,000 species, the beetle family Curculionidae, the weevils, outnumbers all vertebrate animals. Especially in the tropics, many weevils are brightly colored and boldly patterned, like Eupholus bennetti *of New Guinea.*

Opposite. *Many islands are important breeding grounds for bird species such as the red-footed boobies* Sula sula, *pictured here with their neighbors, the frigatebirds, in the background.*

The Hawaiian silversword Argyroxiphium sandwicense *grows only on volcanic ash at very high altitudes. The dense basal rosette and the towering flower stalk are adaptations to extremes of temperature, aridity, and harsh solar radiation—features shared by unrelated high-altitude plants in East Africa and remote areas in the Andes. The plant grows up to 1.8 meters (6 feet) in height.*

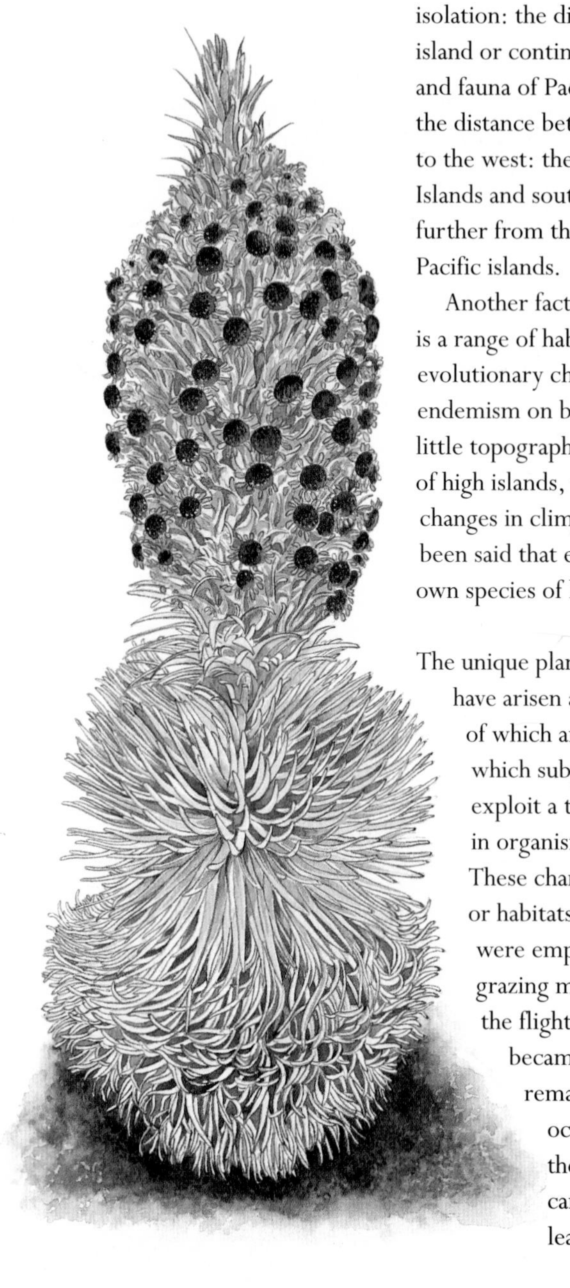

seeds of *Pandanus*, beach morning glory, beach heliotrope, and other plants from the ocean. Seabirds that spend weeks at sea traverse thousands of kilometers of ocean, and the same species—red-footed boobies *Sula sula*, great frigatebirds *Fregata minor*, and white terns *Gygis alba*—are seen in Fiji as on Pitcairn.

Nevertheless, not all seaborne organisms are exactly the same from island to island. Even some of the smallest low islands have their own unique fauna and flora: nine species or varieties of plants, three bird species, and several insects and land snails are endemic to Henderson Island. While ocean currents actively distribute plant seeds and the larvae of marine organisms, some suspension of the means of dispersal must occur, and with it the suspension of gene flow, and subsequent change in organisms with time. Migration is, therefore, not the only factor that explains the peculiarities of Pacific island life.

One of the most effective disruptions to gene flow is isolation: the distance of an island from a neighboring island or continent. Indeed, the diversity of the flora and fauna of Pacific islands is inversely proportional to the distance between an island and the Malayan region to the west: there is more endemism in the Hawaiian Islands and southeastern Polynesian islands, which are further from the Malay Archipelago, than in other Pacific islands.

Another factor which appears to ensure diversity is a range of habitats that provide places in which evolutionary changes can take place. There is little endemism on beaches and low islands, where there is little topographic relief; rather, it is on the inland areas of high islands, with their ridges and valleys, and abrupt changes in climate, that most endemics occur. It has been said that each ridge and valley in Hawaii has its own species of land snail.

ADAPTIVE SHIFTS

The unique plants and animals of the Pacific islands have arisen as a result of species radiations, some of which are recognized as adaptive shifts, by which subpopulations of a successful population exploit a totally new resource or habitat resulting in organisms that diverge far from the ancestor. These changes may occur because the niches or habitats in which the new species developed were empty. For example, in the absence of grazing mammals, reptiles, and other predators, the flightless, ostrich-like moa in New Zealand became a giant grazer. One of the most remarkable of these adaptive shifts has occurred in the Hawaiian inchworms: these herbivorous caterpillars have become carnivorous, feeding on flies, cockroaches, leafhoppers, and spiders.

Sometimes several adaptive shifts have occurred, resulting in a group of organisms occupying a range of habitats. In Hawaii, for instance, the "silversword alliance" comprises 28 species of the endemic genera *Argyroxiphium* (the famous silversword), the *Dubautia* (found on all the main islands), and *Wilkesia* (endemic to Kauai), all derived from a tarweed (Madeiinae) from the Pacific coast of North America.

STEPPING STONES, TIME, AND THE HUMAN IMPACT

Biogeographers read the past to explain present distributions, and three relatively recent discoveries point the way to future understanding of the distribution and evolution of the flora and fauna of Pacific islands.

The discovery in the 1950s of chains of guyots or seamounts, from Australia and Asia to Hawaii, which lent support to the concept of continental drift and plate tectonics, has suggested a means for dispersal by island hopping. These drifting, eroding, and soon-to-be subducted islands, therefore, could have provided stepping stones by means of which the ancestors of Hawaii's plants and animals colonized islands newly arising from the hot spot.

Naturalists have argued that some of the unique plants and animals of Pacific islands are (or were) relics, surviving from times long past. The moa in New Zealand, the iguana in Fiji, and the gymnosperm *Auracaria* and cycads of the western fringe of the Pacific are representative of ancient groups of organisms. Dating of volcanic islands throughout the Pacific indicates, however, that almost all of them are only a few million years old. Indeed, Hawaii is less than 700,000 years old, which implies that the species endemic to it must also be new. There is now every indication that the development of new species on oceanic islands is a very rapid process, and that the time span in which, for example, flightlessness in rails evolved, may be measured in generations, not millennia.

The possibilities of rapid evolution were given impetus in 1970 by the discovery of the bones of a flightless goose on the island of Molokai in the Hawaiian Islands. That discovery has had three significant consequences: the finding of other fossil bird bones, many of them representing both recent and extinct species, throughout the Pacific; the recognition that several of these recent island endemics had ranges far greater than we know them today; and the belief that human impact is the most plausible explanation for their extinctions. The surviving birdlife thus poses the challenge of reconstructing the distribution of land birds on Pacific islands in the past, and thereby continuing an ever-changing understanding of the biogeography of Pacific islands.

E. ALISON KAY

LAND OF THE MOA

JOHN C. YALDWYN

As the processes of continental drift began to separate New Zealand from Gondwana about 100 million years ago New Zealand and its plant and animal life became more isolated. The Tasman Sea, which divides New Zealand from continental Australia, reached its present width about 60 million years ago, and since then, all colonizing animals have arrived by flying, swimming, or floating. While many groups of birds and a few bats have arrived this way, New Zealand is unique in that it is the only sizeable landmass never to have felt the tread of native land mammals.

In many ways the plants of New Zealand, like those of Hawaii, typify the "isolated island syndrome". More than 80 percent of the vascular plants are found nowhere else. Some genera, such as *Coprosoma*, have evolved into many distinct forms as a result of isolation and adaptation to a variety of habitats. A relatively large number of species have male and female plants. New Zealand's flora has a high rate of natural hybridization and, with a few exceptions, its flowers are not notably large or brightly colored.

The presence of conifers, southern beeches, and other woody families not found on other isolated islands, however, clearly demonstrates New Zealand's past links with Gondwana. One of the two main kinds of forest, the conifer broadleaf forest, is an ancient mid-latitude rainforest. In tropical rainforests tree species are numbered in hundreds, and all are flowering plants; in the conifer broadleaf forests of New Zealand, tree species are numbered in dozens, with a mixture of conifers and flowering plants.

The other type of forest present, the beech forest, is dominated by *Nothofagus*, a Southern Hemisphere genus related to the oaks and true beeches of the Northern Hemisphere. This forest type is characterized by a small number of tree species—there are only four *Nothofagus* species in New Zealand; by sparse undergrowth and the absence of vines; and by a deep carpet of leaf litter. The relatively open *Nothofagus* forest developed in both lowland and montane regions in the absence of browsing mammals, and as might be expected, the introduction of deer in the nineteenth century caused vast damage to this distinctive forest type.

There are at least 54 shrub species, in 20 genera and 17 plant families, which grow in a densely interlaced, springy, often cushion-like form, with very small leaves and slender wiry stems. These so-called divaricating shrubs are characteristic of New Zealand vegetation, and

Kim Westerskov/Oxford Scientific Films

New Zealand's spectacular pohutukawa Metrosideros tomentosa *is often known as the Christmas tree for its habit of flowering in December or January.*

Eric Crichton/Bruce Coleman Ltd.

Right. *The four New Zealand beech species of the genus* Nothofagus *dominate much of New Zealand's South Island forests. Their closest relatives are found in eastern Australia and Chile.*

among the reasons suggested for their development are reaction to drought, resistance to wind, frost or desiccation; or the response of an essentially subtropical flora to the onset of glacial climates. However, the most intriguing theory is defense against the browsing of moas—although a divaricating habit would not prevent damage from the shearing action of moa beaks, it would certainly minimize the damage and enable plants to recover quickly.

Polynesians arrived in New Zealand about 1,000 years ago, bringing dogs, rats, and fire—a fatal combination which initiated a series of ecological changes. Extinctions directly linked to the arrival of humans include the entire order of moas, a large goose, a swan, several ducks, a giant eagle, and more than a dozen other birds. This dramatic loss continued unabated with the coming of European settlers and their introduced plants and animals, less than 200 years ago.

Like the finches of Galapagos and the honeycreepers of Hawaii, the moas of New Zealand exhibited spectacular species radiation, developing variations in feeding mechanisms and body size as a result of isolation, changing climate, and modification of environment over time. There were once 11 species of moas in six genera, ranging in size from 1 meter (3 feet) to the giant *Dinornis giganteus* and *Dinornis maximus*, both of which stood at about 3 meters (10 feet) and weighed up to 200 kilograms (440 pounds). Their demise was the result of overhunting and the reduction of their habitat by fire.

Including seabirds, some 150 native and endemic bird species now breed in New Zealand and its outlying islands. The phenomenon of "double invasion", when some species periodically make more than one successful colonization, is a feature of New Zealand ornithology. The three best-known double invasions have produced the distinctive black and pied stilts, the pied *Haematopus ostralegus* and variable oystercatchers *H. unicolor*, and the New Zealand tomtits *Petroica macrocephala* and robins *P. australis*. The endemic black stilt *Himantopus novaezelandiae*, for example, is the product of an early invasion of the Australian pied stilt *H. himantopus*. Now endangered, it has been virtually replaced by a second invasion of pied stilts, possibly as recently as the nineteenth century. •

John Cancalosi/AUSCAPE

Deni Bown/Oxford Scientific Films

Above. *The nuts of southern beeches such as this silver beech* Nothofagus menziesii, *are heavy and unpalatable, and are seldom carried far by wind or animals. As a result, southern beeches have not recolonized many areas of New Zealand that were buried under glaciers during the ice ages.*

Above left. *A survivor from the age of dinosaurs, the tuatara* Sphenodon punctatus *is an extraordinary reptile which is confined to small islands off the New Zealand coast. It has no close relatives anywhere.*

Below. *New Zealand moas probably became extinct about 250 years ago. Eleven species are known to have existed. The giant* Dinornis maximus *is pictured here.*

Frans Lanting/Minden Pictures

The male frigatebird exhibits one of the bird world's most bizarre courtship rituals: It pumps up its enormous balloon-like gular sac, rapidly opens and shuts its mandibles, waves its outstretched wings, and squeals and croaks beseechingly—all to coax females flying overhead to come and mate with him.

The "Enchanted Isles"

Charles Darwin's conclusions from his visit to the Galapagos Islands in 1835 changed scientific thinking about the evolution of life. Today the islands are still natural, but vulnerable, laboratories in which we can study the processes that contribute to the adaptation and diversification of life.

Arising from a volcanically active hot spot in the Earth's mantle, the oldest islands in the Galapagos Archipelago appeared above sea level some three to four million years ago. Since their formation, the islands have remained geographically isolated, never having been much closer to adjacent landmasses than they are today—1,000 kilometers (600 miles) west of the South American continent, and 5,600 kilometers (3,500 miles) east of the nearest Pacific island chain. Terrestrial immigrants colonized them by rafting (on floating debris or on the feathers and feet of birds), by flying, or by wind dispersal. Shallow-water marine organisms arrived either directly by swimming or as minute larvae, carried to the islands on ocean currents.

The difficulties of migration imposed by their watery isolation explain in large part the unique flora and fauna of the Galapagos. The only native land mammals are small rice rats. The dominant fauna on land are reptiles—such as tortoises, lava lizards, marine iguanas, and land iguanas—and birds, particularly small finches, mockingbirds, and one species of hawk. Overall, the Galapagos has a remarkably low diversity of organisms, considering the exceedingly rich assemblages of plants and animals found along adjacent South American shores. There are, for example, only 500 native plant species in the Galapagos compared with more than 10,000 species in Ecuador.

LIMITED LIFE-FORMS

Why this disparity? There are three major reasons. Firstly, Galapagos is geologically quite young and the terrain is dominated by recent lava flows. Only

KINGDOM BY THE SEA

A harsh environment of torrid rock and cold seas, the shores of the Galapagos harbor a wealth of wildlife found nowhere else in the world. From left to right, a bright orange Sally Lightfoot crab scuttles across the rocks; marine iguanas bask in the sun; a fur seal disports in the tidepool; a pair of Galapagos penguins take a break from fishing; a Galapagos mockingbird surveys the scene from its cactus perch; and a flightless cormorant spreads its much-abbreviated wings to dry. Frigatebirds patrol the distant sea.

10 percent of Fernandina, the youngest and most volcanically active island, is covered in vegetation. The arid climate of Galapagos delays the breakdown of rock to soil, leaving much of the shoreline a formidable and hostile environment for both plants and animals. In short, there is a restricted variety of habitats.

The second reason for the limited diversity of life-forms can be attributed to differing abilities to survive transportation to the islands. For organisms coming from the adjacent shorelines (South and Central America being the nearest and most direct source of immigration), transport on a raft under the best of circumstances (a current speed of 1 kilometer/about ½ mile per hour) would require a journey of 42 days. For most organisms, particularly mammals, going without food and water over this length of time would almost certainly be fatal. On the other hand, because of their low metabolic requirements, reptiles have the potential to survive for long periods without sustenance.

The last reason for the low diversity of organisms is chance. With sufficient time, most events will occur. Given the youth of Galapagos, however, there have probably been only a few hundred migrations since the islands formed, and only a small percentage of these would have resulted in the establishment of persistent populations. This conclusion is supported by the results of a recent study of the dispersal of the fruit and seeds of Galapagos plants. In contrast with Hawaii, it was estimated that a minimum of 413 colonization events were required to account for the floral diversity of the islands. Given the maximum age of the islands as five million years, the rate of a successful colonization would be in the order of one species every 12,000 years! This slow rate of colonization, however, is not likely to have been constant. During the earliest stages of island development, colonization may have been even slower; but as conditions for life improved, with the establishment of pioneer species such as lichens and other arid-adapted vegetation, organisms probably colonized more rapidly.

Frans Lanting/Minden Pictures

The marine iguana Amblyrhynchus cristatus *is the world's only sea-going lizard. It is confined to the rocky shores of the Galapagos Islands and feeds on underwater algae.*

Frans Lanting/Minden Pictures

Frans Lanting/Minden Pictures

Stan Osolinski/Oxford Scientific Films

Top. *Otherwise small and dowdy in appearance, the various species of Darwin's finches differ strikingly in bill structure. Seed-eaters, for example, have deep and stout bills, whereas the cactus finch* Geospiza scandens, *pictured, has a moderate and slightly downcurved bill with which it feeds largely on the fruit, pulp, and nectar of the prickly pear.*

Bottom. *Once common across the archipelago, the Galapagos hawk* Buteo galapagoensis *is now much reduced in number, its population perhaps no larger than 300 individuals.*

Right. *The tiny island of Santa Fe is famous for its bizarre forest of giant prickly-pears* Opuntia echios.

ENDEMIC SPECIES

Since successful colonization was rare, most of the species that managed to become established on Galapagos shores became genetically isolated from their parent population. They subsequently evolved independently, responding to a different set of environmental conditions from those in which they originated. Within a very short period of time (believed to be less than one million years) many Galapagos species diverged to the point where they are now quite distinguishable either morphologically (in form and structure) or physiologically (in their functioning) from their ancestors. About 34 percent of the terrestrial plants, more than 50 percent of the terrestrial animals, and about 25 percent of the marine flora and fauna are endemic. Particularly remarkable is the fact that many of the founding species eventually migrated to other islands within the archipelago and further diverged into a variety of distinct forms—some clearly differentiated into separate species, others only slightly modified in appearance and thus accorded the status of subspecies or race.

The diversity of form within a particular group, such as the tortoises or finches, from what is believed to be a single common ancestor is known as species radiation. Examples of species radiation occur within Galapagos in nearly every major terrestrial group of organisms—tortoises, iguanas, lizards, land snails, and prickly pear cactus among others. The most famous example is Darwin's finches, with 13 recognized species in Galapagos and one on Cocos Island, some 540 kilometers (335 miles) to the north.

DARWIN AND THE GALAPAGOS

The unique flora and fauna on islands adjacent to the rich and distinct biota of South America, and the variation between populations of mockingbirds, tortoises, and plants at several localities within the Galapagos Archipelago, provided Charles Darwin in 1835 with evidence that species were not fixed, but mutable (that is, able to undergo morphological change). These observations supported his hypothesis that species are able to change in response to different environmental conditions, and provided the foundation for the development of the theory of evolution based on the principle of natural selection.

While Darwin was correct about natural selection, he did not recognize the importance of isolation as a requisite for the divergence of species. This is perhaps not surprising: at the time his ideas were formulated, Darwin was unaware of genetics, and did not know that interbreeding populations would dampen the effects of selection on the population as a whole.

The importance of isolation, however, was recognized even during Darwin's time. In 1868, only nine years after the publication of *The Origin of Species*, the German zoologist Mortiz Wagner wrote: "The formation of a real variety which Mr. Darwin considers as incipient species, can succeed in nature only where some individuals can cross the previous borders of their range and segregate themselves in nature for a long period from other members of their species."

These early discoveries in Galapagos made a significant contribution to the development of evolutionary theory, and the islands continue to provide scientists with valuable information about the ecological factors that shape evolution. For example, recent studies of Darwin's finches have revealed that

DARWIN'S FINCHES

The Galapagos finches were the bird species that may have been the starting point for the development of Darwin's famous theory of evolution. Darwin's finches, which now occupy most islands and most habitats in the Galapagos Islands, evolved from a single ancestral species.

Tui de Roy/AUSCAPE

Tool-using was once considered a uniquely human ability, a belief that was shattered when it was discovered that the Galapagos woodpecker finch Cactospiza pallida *habitually uses cactus spines as tools to extract grubs from bark crevices. Many other animals have since been identified as tool-users.*

Below. *Giant tortoises* Geochelone elephantophus *on the slopes of Alcedo Volcano, Isabella Island.*

Opposite, top. *The Galapagos marine iguana* Amblyrhynchus cristatus *thrives on marine algae along the volcanic shoreline. It can dive to 10 meters (30 feet).*

Opposite, center. *The Galapagos harbors a wealth of reptiles. Here a snake strangles a lava lizard, which is a smaller cousin of the iguanas.*

Opposite, bottom. *A giant Galapagos tortoise at lunch. In the days of sail, mariners quickly discovered that these ungainly beasts could be kept alive on board ship for up to 14 months, and it became common practice to call in at the archipelago to stock up on "Galapagos mutton".*

food limitation is the driving force that determines bill shape and feeding behavior. These morphological and behavioral adaptations, in turn, influence mating behavior, which leads to reproductive isolation and reinforcement of species diversity. Darwin's finches have evolved in response to alterations in their physical environment and to indirect changes caused by competitors.

THE MARINE ENVIRONMENT

The Galapagos Islands have also provided a good opportunity to study the evolution of adaptations in the marine environment. Situated at the confluence of a dynamic current system, the islands are seasonally bathed by cool nutrient-rich waters and then by warm nutrient-poor waters. The cool waters have enabled temperate-climate organisms such as fur seals, sea lions, cormorants, and penguins to colonize. However, the warm waters that prevail over four to five months of the year pose potential problems in terms of food supply and overheating.

One adaptation to minimize overheating is a reduction in body size: the Galapagos fur seal is the smallest marine mammal, and the penguin is the second smallest of its kind. In contrast with their temperate-water cousins, Galapagos sea lions forage mainly during daylight hours and mate in the water rather than on land. Both of these behavior patterns are believed to represent adaptations that minimize exposure to the tropical heat.

MARINE IGUANAS

The Galapagos marine iguana *Amblyrhynchus cristatus* has no counterpart anywhere else in the world. With the exception of a few species of lizards known to forage on seashore invertebrates, it is the only truly semi-aquatic marine lizard. Thought to share a common ancestor with the land iguana, the marine iguana has adapted morphologically for feeding in the sea.

These somber animals often occur in large numbers (up to several hundred), and may attain a length of 1.2 meters (4 feet) and a weight of 12.5 kilograms

D. Parer & E. Parer-Cook/AUSCAPE

(27½ pounds). Its tail is laterally compressed to help in swimming; it has a blunt snout to facilitate grazing on the turf-like algae; and its long sharp claws enable it to cling tightly to rocks, thereby avoiding being swept away or battered on the rocks by the surf. The Galapagos population of marine iguanas is conservatively estimated to number 300,000 individuals.

Marine iguanas are well known for their ability to swim offshore and dive to 10 meters (30 feet) deep to feed on algae covering the sea floor. Such forays usually last from 60 to 90 minutes with 2 to 15 minutes spent submerged. Feeding in cool water (15–25°C/59–77°F) can pose energetic problems for ectothermic (cold-blooded) creatures such as iguanas, whose peak activity levels occur when their body temperature is at 35°C (95°F). To achieve this temperature, iguanas assume a basking posture with their appendages fully extended and bodies pressed against the islands' lava rock. Their black body coloration and the high heat-radiating potential of the rock enable them to reach their ideal body temperature, even in the early morning in overcast conditions. Once sufficiently warmed, an iguana is able to venture into the sea and stay there until its body cools to about 25°C (77°F).

Because smaller animals cool faster than larger ones, the ability to forage offshore is limited by the size of the individual. Below a certain body mass the amount of food (energy) gained by offshore feeding is minimal compared to the energy (swimming) expended to obtain it. Thus, two different feeding strategies have evolved: juveniles and smaller-sized females weighing less than 1.2 kilograms (2½ pounds) feed on algae found in the intertidal zone, while individuals larger than 1.8 kilograms (4 pounds) are able to feed in subtidal habitats.

Physiologically, marine iguanas have apparently adapted very little to their aquatic existence. Their only clear adaptation is their nasal salt gland, which has the highest secretion rate of any reptile and which allows them to feed on salt-rich algae with no source of drinking water other than the sea.

Since marine iguanas spend only 5 percent of their daily activity feeding in the sea, it has been argued that natural selection has favored physiological attributes that maximize performance in the social and mating activities that take place on land, where air temperatures are high.

The reason that marine iguanas evolved in Galapagos and are found only there is speculative, but is likely to be explained by a number of factors. Firstly, much of the coastline of the islands is desolate and arid but food can be found at the shoreline. Secondly, conditions for a marine lifestyle are particularly suitable in this unusual

Mark Jones/AUSCAPE

Frans Lanting/Minden Pictures

Frans Lanting/Minden Pictures

Frans Lanting/Minden Pictures

Closely related to the marine iguana but differing most obviously in its longer snout, a land iguana Conolophus subcristatus *shelters beneath its favorite food-plant,* Opuntia.

tropical location, where a cool, marine climate promotes rich algal growth, and the land is consistently warm and conducive to an ectothermic existence. Lastly, the absence of native, land-based predators, such as cats and dogs, has permitted the iguana to lead an otherwise vulnerable lifestyle.

SCIENCE AND CONSERVATION

There is still much to learn from the Galapagos life-forms. Newly developed biochemical techniques are currently being applied to reveal the genealogy of various Galapagos species within the archipelago and identify their mainland ancestors. Knowledge of the time when species diverged, coupled with data on geological and climatic changes in the islands, will yield estimates of the rates of evolution and give us some idea of the environmental conditions that may influence this change.

While the Galapagos Islands are well known for their contribution to our understanding of evolutionary processes, they have also been recognized as a model for wildlife conservation. Organisms native to island ecosystems have evolved differently from their mainland counterparts, and as a result are often negatively affected by introduced species. The most serious threat to Galapagos wildlife is not tourism or human habitation, but rather introduced animals such as goats, pigs, cats, dogs, and rats; and introduced plant species such as trees and vines.

Initial steps to safeguard native species were taken in 1934, when the government of Ecuador provided protection for certain animals and made legal provision for the declaration of reserves and national parks. In 1957, at the request of the Ecuadorian government, the United Nations Educational, Scientific and Cultural

Dieter & Mary Plage/Survival Anglia

Organization (UNESCO) and the World Conservation Union (IUCN) commissioned a study to determine the status of the unique wildlife in Galapagos. During the next seven years, an international scientific organization was formed, and the Charles Darwin Research Station (CDRS) was constructed on the island of Santa Cruz. The CDRS defined the boundaries for reserves and conducted programs to control or exterminate feral species that were damaging the environment. By 1968, 97 percent of the landmass in Galapagos was given complete protection as a national park.

This alliance, between a national government and an international science organization, has been extremely

With the sea as its only source of water, the marine iguana has had to evolve effective salt-processing equipment in the form of special glands in the face that accumulate brine, to be vented from the nostrils in sneeze-like showers.

successful. Goats have been eradicated from five islands. In the absence of predators, populations of goats had reached several thousand on some islands, leading to the near-extinction of endemic plant species and the loss of food for native grazers such as giant tortoises.

On islands where the giant tortoise population had been decimated by whaling in the eighteenth century, or where introduced black rats preyed heavily on hatchling tortoises, captive breeding and repatriation programs were instituted. This project has been highly successful in stemming the extinction of at least two races of giant tortoises. In 1986 the entire interior seas of the marine environment, consisting of 7 million hectares (17 million acres), was designated a marine reserve, thus ensuring the protection of all marine species and the critical marine habitats of semi-aquatic species such as sea lions, penguins, and cormorants.

The joint effort between government and international organizations to publicize the economic and scientific value of the Galapagos Islands has thus far proven successful in gaining financial and political support to continue conservation work still needed in the islands. As a tribute to this endeavor, in 1980 UNESCO declared the Galapagos Islands a World Natural Heritage site. ■

G.M. WELLINGTON

7 Islands of the Atlantic and Indian Oceans

STORRS L. OLSON AND DIANA WALKER

The islands of the Atlantic and Indian oceans are far fewer and more isolated than those of the Pacific. At first glance their plants and animals seem much less exotic, but recent fossil evidence reveals how numerous and diverse the life-forms of the islands once were. Since human habitation, many species have become extinct and many more are endangered.

Hermann Brehm/Bruce Coleman Ltd.

The Dominican gull Larus dominicanus *breeds on most islands of the subantarctic region, including South Georgia and the Falkland Islands.*

The Atlantic Islands

Atlantic islands such as Newfoundland and Britain were connected to nearby continental landmasses during the ice ages, and the flora and fauna of these continental islands are, thus, remnants of a larger set of species occurring on the adjacent continents. Many Atlantic islands, however, are oceanic islands whose native fauna and flora must have crossed over water to colonize them.

CONTINENTAL LINKS

With the exception of wave-washed St. Paul Rocks, home to but three species of seabird and a few insects, all the oceanic islands of the Atlantic are volcanic in origin. Many are, or were, formed on the Mid-Atlantic Ridge, the underwater mountain range that marks the zone where North and South America pulled away from Europe and Africa, to begin forming the Atlantic Ocean more than 130 million years ago.

As a rule, the flora and fauna of the Atlantic islands are derived from their nearest continent. Thus, the affinities of Bermuda are with North America; the islands of Macronesia (Azores, Madeira, Canaries, and Cape Verdes) are with southern Europe and northern Africa; Fernando de Noronha and Ilha da Trindãde are with Brazil; and Ascension and St. Helena are with Africa. The exceedingly remote islands of the Tristan da Cunha group and Gough Island are interesting because, although they are closer to Africa, their land birds, at least, seem more similar to those of South America. Presumably, this is because the prevailing winds are from the west, so that even today most of the vagrant birds arriving at Tristan da Cunha are of American origin.

The natural history of islands at the northern and southern extremes of the Atlantic reflects their proximity to the poles. The wildlife of Iceland, for example, is like that of northern and Arctic Europe, whereas South Georgia is populated mainly by Antarctic seabirds, and the Falklands (Malvinas) by a combination of Antarctic and South American groups.

A string of volcanoes in the Gulf of Guinea gave rise to the islands of Bioko (Fernando Póo), Príncipe, São

Michael Dixon/C.M. Dixon

Tomé, and Annobón (Pagalu). Bioko is continental, but the others are oceanic and support several species that occur nowhere else. These endemic species include a dwarf ibis *Bostrychia bocagei*, a pigeon *Columba thomensis*, a very large weaverbird *Ploceus grandis*, and a grosbeak *Neospiza concolor* with a truly massive bill. This last species is now presumed extinct, and with the destruction of the museum in Lisbon by fire, the only remaining specimen appears to be that in the British Museum. The overall affinity of the flora and fauna of these islands is overwhelmingly with that of equatorial Africa.

HUMAN SHADOW

In the north Atlantic, with the exception of Bermuda, the oceanic islands are all part of larger archipelagos. There is no prehistoric record of human occupation of any of these islands except the Canaries, which were colonized by a people known as the Guanches between 2500 BC and 2000 BC. Most of the other islands and archipelagos were discovered and inhabited during the era of great Portuguese and Spanish exploration in the fifteenth and early sixteenth centuries.

Human settlement has had a similar impact on the ecology of all the Atlantic islands. Historical accounts

The rugged volcanic cliffs of Ponta de Sao Lourenco at the eastern tip of Madeira. Uninhabited and densely wooded when discovered in 1419, the island's forests have been extensively destroyed by fire during subsequent settlement.

Michael Brooke/Oxford Scientific Films

Few islands are so distant from the nearest land that they have never been colonized by some bird. Even remote Inaccessible Island in the south Atlantic has its unique species, the rail Atlantisia rogersi.

The skeleton of a baby rail, with its relatively tiny wings and feeble breastbone. These structures are far larger in flying adults.

invariably mention the release of destructive domesticated animals. Goats and rats were among the first, and were usually followed by pigs, dogs, cats, mice, and other pests that either killed native animals directly, or destroyed the vegetation on which they depended. On those islands that were settled by Europeans, habitat destruction continued through burning, clearing for agriculture, and the introduction of noxious plants and insects.

Biologists did not come to study the islands until long after these perturbations had altered the character of native ecosystems. Thus, the full extent of human impact was not appreciated until systematic searches were made for fossils of extinct vertebrates, beginning in the 1970s. These studies revealed many extinct or extirpated populations of seabirds, land birds, a few reptiles and mammals, and land snails. So far, the search for fossils has covered Bermuda, Madeira and Porto Santo, the Canaries, Fernando de Noronha, Ilha da Trindãde, Ascension, and St. Helena. Extinct forms have been found on all but Trindãde, where the geological environment is not conducive to fossil deposition.

While many of these studies are still incomplete and unpublished, they all show that the historically known fauna of these islands is often only a pitiful remnant of what was there before the arrival of humans. There are no endemic species of land birds on Bermuda now, yet in the past there was a finch, a woodpecker, an owl, a heron, and several flightless rails. The most distinctive land birds of Madeira are a species of pigeon and a pipit that also inhabit the Canaries. But fossils from Madeira have also revealed extinct thrushes, finches, a rail, and most interesting of all, two species of quail that may have been flightless. Extinct rodents, lizards, a tortoise, and a finch have been described from fossils in the Canaries. With this in mind, the absence of any endemic species of bird in the Azores seems quite improbable. These islands are older and more remote than the Galapagos, and it is hardly conceivable that no differentiation of birds beyond the level of subspecies took place here.

FOSSIL EVIDENCE

The realization that recent extinctions have occurred so commonly on islands should make us view anomalous patterns of distribution with skepticism. For example, although there are 10 much larger islands in the Cape Verde archipelago, the lark *Alauda razae* occurs only on the tiny barren islet of Razo, and nowhere else in the world. Likewise, the impressive giant skink *Macroscincus coctei*, a lizard over 30 centimeters (1 foot) long not including the tail, is confined to Razo and the adjacent, even smaller, islet of Branco. These distinctive organisms surely did not evolve in response to the conditions presented by these tiny islets. It is more reasonable to think, given the evidence on extinctions, that they must once have been more widely distributed in the archipelago, but persisted on Razo and Branco only because these islands are barren, waterless, and, therefore, uninhabited by humans.

Rails and gallinules are somewhat chicken-like marsh birds that disperse widely and are often found as vagrants far out of their normal range. For this reason they are very successful at colonizing remote oceanic islands, where they have often quickly evolved flightlessness. This condition unfortunately renders them extremely susceptible to introduced predators. The only flightless members of this group known historically in the Atlantic are gallinules from Tristan (now extinct) and Gough islands, and the tiny flightless rail of Inaccessible Island, all in the Tristan da Cunha group. But fossils show that there were probably flightless rails on all the islands of the Atlantic, for in addition to those mentioned above from the north Atlantic, we now know of a flightless rail from Fernando de Noronha, two from St. Helena, and one from the harsh and inhospitable island of Ascension, which is hardly more than a giant cinder. This rail was observed alive and accurately described in 1656 by the astute traveler and diarist Peter Mundy. It is thought to have been dependent on carrion and associated insects found in seabird colonies, and indicates the extreme adaptability of rails to various island environments.

One of the more peculiar discoveries on an Atlantic island were bones of a hoopoe on St. Helena. The hoopoe *Upupa epops* is an odd bird with a long bill and a striking crest that occurs throughout Europe, Africa, and Asia. The bird on St. Helena was, however, a distinct species. It was considerably larger in size and had much smaller wings than mainland hoopoes. Although it was almost certainly not flightless, it was on its way to becoming so. A flightless hoopoe could scarcely be imagined were it not for the bones that remained on St. Helena.

FLIGHTLESSNESS

Because the evolution of flight was the key adaptation that allowed birds to diverge and radiate from their reptilian ancestors, it is important to ask how and why birds so often become flightless in island environments. In the Atlantic this condition is confined almost entirely to rails, but elsewhere we know of flightless ducks, geese, ibises, pigeons (the dodo of Mauritius), the kagu of New Caledonia, and other groups of insular birds that have become flightless. One of the principal factors that allows flightlessness to evolve on islands is the almost invariable absence of mammalian predators. It is thought that very simple changes in the process of development explain the evolutionary rapidity with which flightlessness occurs. At hatching, all rails, for example, have large, well-developed legs and feet, but tiny wings and a very reduced breastbone. Only later in development does the flight apparatus increase in relative size to permit the bird to fly. By arresting the development of the flight apparatus and retaining the proportions of a chick into adulthood, a bird would thus remain flightless.

This explains how a bird may become flightless. But why? The answer lies in the conservation of energy. Flight has a great energy cost. It requires large breast muscles and an elevated metabolic rate for those muscles to function properly. Therefore, when conditions permit it, if large breast muscles and a higher metabolic rate can be eliminated, a bird will have greatly reduced energy requirements. Less food would be needed to sustain the individual and, therefore, more energy would be available for egg-laying and other reproductive activity. Recent studies have confirmed that flightless rails do indeed have lower metabolic rates than their closest flying relatives. Thus, the evolutionary advantages of flightlessness in an insular situation are very important, and we find that flightlessness has evolved repeatedly in different kinds of birds the world over. Flightless birds, however, are immediately at an extreme disadvantage once humans introduce unnatural predators into their untroubled homes. Thus, the true prevalence of this adaptation has been revealed only recently through the fossil record.

CLIMATE CHANGE

Not all extinctions on islands are attributable to humankind. The repeated advances and retreats of glaciers in the past million years or so caused wide fluctuations in sea levels and profound changes in marine environments. Bermuda, for example, shrank to around a tenth its present size when the seas rose to their maximum extent some hundreds of thousands of years ago, but at the height of the last glaciation, about 18,000 years ago, when sea levels dropped precipitously, Bermuda was 10 times larger than it is now. At that time, we know from fossils, a crane, a duck, and four different species of flightless rails existed on Bermuda that have since disappeared. A few of these doubtless survived the subsequent reduction in land area, but the inundation that reduced Bermuda to its present size would surely have eliminated all the habitat available for a crane.

On St. Helena, a pigeon and a shearwater related to the Indo-Pacific wedge-tailed shearwater *Puffinus pacificus* are known only from the oldest deposits on the island. These birds presumably disappeared due to natural changes in the environment, before the arrival of humans. Nevertheless, these documented instances of "turnover" unrelated to human activity are very rare.

A map of Bermuda showing the present and past extent of the island. During the last ice age, the central lagoon was probably a large marshy area suitable for cranes, which once inhabited the islands but became extinct when sea levels rose.

Graham Robertson/AUSCAPE

Like many related seabirds, the soft-plumaged petrel Pterodroma mollis *lays a single egg at the end of a burrow dug in the soil, which it visits only at night. This species breeds commonly on Gough Island, south Atlantic Ocean.*

L.A. Fuertes/Robert Harding Picture Library

The last recorded specimen of the Labrador duck Camptorhynchus labradorius *was shot on Long Island, New York, in 1875. Although the reasons for its extinction remain somewhat mysterious, human persecution is unlikely to have been a direct cause, as the bird was notoriously shy and not particularly good to eat.*

DISAPPEARING SEABIRDS

Although terrestrial organisms were hard hit by the arrival of humans, seabirds were dealt perhaps a more devastating blow. Even the most oceanic of birds must come to land to nest. Thus, for seabirds, the few tiny islands of the Atlantic were of inestimable importance, far out of proportion to their total land area.

The most renowned of seabirds dependent upon Atlantic islands was the flightless great auk *Pinguinus impennis*, an ecological counterpart to the flightless penguins of the Southern Hemisphere. This bird is known to have nested in historic times on islands from the Gulf of St. Lawrence to the British Isles. In prehistoric times it was known from Florida to the Mediterranean, so the species had probably been adversely affected by prehistoric people before Europeans exterminated it for food and oil. The last known pair was killed in 1844.

Another extinction in the north Atlantic that has never been satisfactorily explained is that of the Labrador duck *Camptorhynchus labradorius*. This beautiful sea-duck was known only on its wintering grounds from New Brunswick to New Jersey in North America. It was never common, and although shot for the market, it was not considered good to eat, so hunting was probably not a factor in its initial rarity. The species is known from fewer than 50 specimens, and disappeared about 1875. Perhaps this duck also nested only on islands, where prehistoric exploitation may have reduced its numbers long before the species was discovered and named by naturalists.

Since it surfaced some 30 million years ago, Bermuda probably has been an important breeding locality for seabirds. The petrel known as the cahow *Pterodroma cahow* once occurred there by the millions, but the species was reduced to a pitiful few dozen individuals by colonists, who used it as food, and by pigs, cats, dogs, and rats. Fossils show that St. Helena, too, was once home to vast numbers of seabirds, at least six species of which were exterminated after human settlement. The few remaining species exist only in very low numbers on small offshore rocks.

With seabirds the story is not so much one of complete eradication of species, although this did happen, but of reduction of population sizes by several orders of magnitude. Millions of petrels, terns, frigatebirds, boobies, and other efficient surface predators were removed from the oceanic environment. What the effect may have been on the fish and squid that were their prey, or on nutrient recycling in the waters around the islands where they nested, has never been calculated.

Erwin & Peggy Bauer/Bruce Coleman Ltd.

J.A.L. Cook/Oxford Scientific Films

Common across North America, the yellow-crowned night heron (far left) has recently been reintroduced to Bermuda in an attempt to restore a vanished natural system of checks and balances: once the night heron was the chief predator of a local land crab Gecarcinus *(left), which preys heavily on chicks of the now critically endangered Bermuda cahow.*

RESTORING THE BALANCE

The extent of previous destruction, coupled with the remoteness of many of the Atlantic islands, would at first seem to make conservation measures a dismal enterprise. There are some reasons for hope, however. Heightened environmental awareness, especially on the more populated islands such as Madeira and the Canaries, should help with efforts to preserve what remains of the fauna and flora. More tangibly, on St. Helena there has been a vigorous effort to identify and preserve, through propagation and reintroduction, the nearly vanished native vegetation, which includes many highly peculiar plants that are thought to be of relatively ancient origin.

Finally, the determined efforts of David Wingate, Conservation Officer on Bermuda, should stand as an example of the effect of individual perseverance and optimism in the face of the bleakest of prospects. Bermuda is one of the most densely populated islands in the world, besides having more than 7 percent of its land area taken up in golf courses. By the first part of this century, very little remained of its native vegetation and what did was dealt a devastating blow by the introduction of a scale blight from California in 1942, which denuded the landscape of cedar forests within a few years.

The one endemic species of bird on Bermuda, the cahow (once thought to be extinct), persisted only as a few pairs on tiny offshore islets. But through Wingate's constant monitoring and creative environmental manipulation since the 1960s, the cahow has slowly been brought back from the brink of extinction. Its numbers have continued to increase despite setbacks such as the appearance of a vagrant snowy owl that killed several breeding birds. One factor that had kept numbers low was the insufficiency of suitable nesting sites, but Wingate cut artificial nest burrows out of solid limestone, and the birds readily adopted them. Another limiting factor is predation on chicks by land crabs. Although land crabs occur naturally in Bermuda, we know from fossils that one of their chief predators, a form of yellow-crowned night heron with an exceptionally large bill, had been exterminated. As a remedy, yellow-crowned night herons were introduced from the North American mainland and have started breeding on Bermuda, where it is hoped they will keep the number of land crabs in check.

One of Bermuda's larger islets, Nonsuch, has been set aside as a nature reserve. Through intensive effort, most alien animals and plants have been removed from Nonsuch and native plants, including cedar, carefully nurtured. In a few more years, it is hoped that Nonsuch will be restored to a fairly faithful approximation of what all of Bermuda was like before human settlement. Perhaps then the cahow will reclaim the island for itself, showing that it is possible to undo at least some of the damage that humans have wrought in their conquest of the Atlantic islands.

STORRS L. OLSON

A large flightless seabird, the extinct great auk Pinguinus impennis *inhabited islands off northern Europe, Greenland, and northeastern Canada. The last were killed by a fisherman on 3 June 1844.*

Max Gibbs/Oxford Scientific Films

Common, colorful, and widespread, parrotfish are mostly associated with coral reefs in the Caribbean, Pacific, and Indian oceans. This is the clown parrotfish Cetoscarus bicolor.

Opposite. *A beach on Mahé, Seychelles. Mahé and its neighbors Praslin, La Digue, and Silhouette are the only sizeable granitic oceanic islands in the world.*

Land crabs inhabit forest but must return to the sea to breed. In the case of the red crab Gecarcoidea natalis, *which is endemic to Christmas Island, this movement takes the form of a spectacular annual migration.*

Islands of the Indian Ocean

Like the islands of the Atlantic, Indian Ocean islands are not as numerous as those of the Pacific. Nevertheless, they are more varied, with many unique and interesting aspects. Relatively few of them are inhabited, but many have suffered from human exploitation.

CORAL TO RAINFOREST

Archipelagos of many small coral atolls are rare in the Indian Ocean—the Maldives and Laccadives, and the Chagos Archipelago, are the two main groups. However, there are many single coral atolls, as well as much larger islands such as Madagascar and Sri Lanka. Off the northern coast of Western Australia the several isolated coral atolls known as the Rowley Shoals are composed of three atolls, and Scott, Ashmore and Seringapatam reefs. They have little landmass above the water and are uninhabited, and although there is little terrestrial scenery, their subtidal environments are often spectacular, with extensive coral reefs, steep drop-offs, and large territorial reef fish. The northern reefs support very dense sea-snake populations, making diving an interesting experience!

Whereas these islands are low and bare, Christmas Island, 400 kilometers (240 miles) south of the Indonesian island of Java and part of the same underwater chain of volcanoes as the Cocos (Keeling) Islands, rises to a height of 361 meters (1,185 feet). With an area of 13,700 hectares (33,840 acres), it provides a range of habitats from fringing coral reef to rainforest.

The higher part of the island known as the plateau is covered in rainforest. The rainforest canopy, although similar in appearance to other such canopies, contains only about 10 major species. Some species found as typical tall rainforest trees elsewhere grow as small- to medium-size trees near the shore of the island. Others that would normally occur only as shortlived secondary growth in a rainforest, on Christmas Island form a long-lived part of the canopy. There are many associated plants such as orchids within the rainforest and about 20 to 30 percent of the plant species are endemic.

Christmas Island is a major bird habitat and has several endemic species, including Abbotts booby *Sula abbotti*, the Christmas Island frigatebird *Fregata andrewsi*, and the golden bosun *Phaethon lepturus fulvus*. There is also a large fruit pigeon, the imperial pigeon *Ducula whartoni*, which lives in the rainforest. The limestone of the island is riddled with caves, and the Christmas Island glossy cave swiftlet *Collocalia esculenta natalis* forms its nests high on their roofs.

Two species of bat are also endemic to Christmas Island: the Christmas Island insectivorous bat *Pipistrellus murrayi* and the Christmas Island fruit bat *Pteropus natalis*. The insectivorous bat is very small, with a wingspan less than 10 centimeters (4 inches), in contrast with the fruit bat, which has a 50 centimeter (20 inch) wingspan. These bats are generally associated with the rainforest.

The island is particularly famed for its large land crabs—there are more than 20 species—which are thought to be able to thrive because of the absence of large predators on the island. Three species are most common: the robber crab *Birgus latro*, the blue crab *Cardisoma hirtipes*, and the red crab *Gecarcoidea natalis*. The red crab is ubiquitous on the island, and it has been estimated that there are 120 million individuals with a total weight of 8,000 tonnes! At these densities (more than 1 per square meter), there is little debris in the forests or gardens that the crabs have not removed.

Although land crabs live successfully on land, they have to return to water for the early stages of their development, and some must remain near water. Blue crabs, for example, are not tolerant of very dry conditions even as adults, and are usually found near freshwater streams, where they breed. Other land crabs, such as the red crab, migrate back to the sea en masse at the end of the dry season. Among the robber crabs, only the female migrates back to the sea, having already mated before beginning her journey. This migration is unsuccessful for many years and the few baby crabs that emerge migrate inland when they are only a few weeks old. Despite the poor years, the population level appears to be maintained by an occasional very successful year.

Jan Aldenhoven/AUSCAPE

Jean-Philippe Varin/AUSCAPE

The giant land tortoise Geochelone gigantea *once occurred on Mauritius, the Seychelles, and several other islands in the western Indian Ocean, but by 1840 it was extinct everywhere but Aldabra. Given complete protection in 1891, its population now numbers about 150,000.*

Opposite. *Three Madagascan plants (from top to bottom): the baobab* Adansonia madagascariensis, *often called the monkey-bread tree; the octopus tree, one of an endemic family Didieraceae of 11 species; and* Kalanchoe gastonisboniri *from the rugged Bemaraha plateau of southwestern Madagascar.*

The dodo flourished on Mauritius until the island was discovered around 1507. Tame, flightless, and easy to kill, the dodo was exterminated by 1680.

ISLAND REFUGES

Where islands are remote from major landmasses or other islands, there is less chance of arrival of new species and a greater chance of extinction. Thus, islands such as the Seychelles develop an assemblage of species that may not exist anywhere else.

The Seychelles, which include the Aldabra group, are mid-ocean islands about 930 kilometers (580 miles) north of Madagascar. The Seychelles are not typical coral atolls, but are composed of the continental rock granite. The largest island, Mahé, has an area of 145 square kilometers (56 square miles), but rises to a height of 900 meters (3,000 feet). There are some 40 other islands and islets in the Seychelles, some of which are low sand cays (less than 5 meters/16 feet high) on sea-level coral reefs. Others, including the Aldabra group, are of reef limestone, raised slightly above sea level to a height of about 8 meters (26 feet). The rainfall over these atolls is variable, determined by the elevation of the islands: the large granitic islands have up to five times as much rain as low islands such as Aldabra.

The variation in topography, rainfall, and use of the land has resulted in a mosaic of vegetation, ranging from rainforest to dry scrub. The flora of the Seychelles, like that of Madagascar, shows greater similarities to the western Pacific than to Africa, probably as a result of the arrival of airborne or seaborne seeds and fruits from there. In the eighteenth century coconuts were planted as a cash crop, and with other introduced species such as pineapples, vanilla, and cinnamon had a major destructive effect on the native flora. The tall, lowland forest which once dominated the original flora of the granitic islands, now occurs only in inaccessible inland and upland places.

Unlike larger islands such as Madagascar, the small islands of the Seychelles have few native land mammals. The size and scale of the habitats are apparently too small to allow the survival of even small mammals, although there are two species of bats on the Seychelles and three on Aldabra. In contrast with the low diversity of native mammals, avian fauna has successfully exploited the different habitats available. Fifteen endemic species of birds have evolved, including one of the world's rarest falcons, the Seychelles kestrel *Falco araea*. Of the 40 species of birds on the islands, about 55 percent have affinities to species on Madagascar. There are also many endemic amphibians, including the most unusual legless burrowing caecilian *Hypogeophis*, which cannot swim. Its presence on the Seychelles, while absent from adjacent Madagascar, is difficult to explain.

Farthest to the northwest lies Aldabra, and it has a somewhat different mix of fauna. Aldabra is a low coral atoll of four main islands around a central lagoon, with an area of about 100 square kilometers (40 square miles). It has about a dozen species of endemic land birds, one of which, the white-throated rail *Dryolimnas cuvieri*, is the only flightless bird left in the Indian Ocean. Aldabra is also the only remaining location in the Indian Ocean of the giant land tortoise *Geochelone gigantea*, which was once widely distributed, especially on Madagascar and the Seychelles. These animals can weigh more than 50 kilograms (110 pounds) each, and are slightly larger than their Pacific counterparts on the Galapagos Archipelago. Like the Galapagos tortoises, they were used as food by sailors on long ocean voyages, and by 1840 were extinct, except for those left on Aldabra. After the first settlement was established on Aldabra in 1891, the killing of giant tortoises was prohibited. The population was slow to recover from the effects of exploitation: by 1916 the numbers were still only in the thousands; however, there are now more than 150,000.

Aldabra is not ideal as the last refuge of the giant Indian Ocean tortoise. The islands have low rainfall and a dry season lasting eight months, and the tortoises, which are cold-blooded and cannot regulate their body temperature, are limited by lack of shade, food, nesting sites, and water. As they congregate where there is shade and water, their distribution corresponds with areas of dense vegetation. In areas where no shade is available, chances of survival are slim. In some places there may be more than 70 tortoises per hectare (28 per acre). Fortunately, their future now seems more secure as the islands have become a focus for wildlife conservation. Nevertheless, their conservation requires attention not only to the animals themselves, but also to the vegetation they require for food, shade, and shelter.

Madagascar

Madagascar lies only 400 kilometers (250 miles) off the southeast coast of Africa. It is the fourth largest island in the world, and has been separated from the African continent for perhaps 70 million years. Despite its proximity to the African coast, Madagascar's flora is more similar to the western Pacific than to Africa. It is a very rich and diverse flora with a high degree of endemism—nearly 90 percent of its 7,300 species occur nowhere else in the world.

RICHNESS AND DIVERSITY

Dense rainforest, typical of tropical rainforests in appearance, occurs where the annual rainfall is greater than 1,500 millimeters (58 inches), on the narrow coastal plain to the east of the island and the high plateaus of the interior. On the western side, there is less rainfall from the trade winds, but here the island is subject to monsoonal rains, which decrease to the south. Receiving less than 300 millimeters (12 inches) per year, the extreme south is very arid, with a dry season lasting 10 months. The difference in rainfall between east and west has left the eastern side with 500 genera and 5,500 species, and the western side much barer with only 200 genera and 1,800 species. The two sides have 600 species in common. As well as rainforest, Madagascar's western side has species adapted to the drier climate, including many succulents, and some strange-looking plants that grow on dry rocks, such as *Diderea madagascariensis* and *Pachypodium rosulatum*. There are also extensive grasslands of few native species, but often with many introduced ones. These grasslands are thought to have come about because of the clearing of forests and shrubs for agriculture. Only about 21 percent of the island remains as forest.

PRESENCE AND ABSENCE

Madagascar has an unusual balance and range of fauna. For example, it has over 150 species of frogs, but no toads, poisonous snakes, or freshwater fish. In comparison to the rich avian fauna of southern Africa, the bird population is impoverished with only 82 species, half of which are endemic.

The island's endemic species are often relatively small in size (less than 1 meter/3 feet tall). The most famous of these is the lemur, a large-eyed, primitive primate, which makes up about 40 percent of all mammals on Madagascar. There are more than 20 distinct species of lemurs, ranging from small, mouse-like forms to one large, ape-like form. These animals usually live in the trees on the wet eastern side of Madagascar, but the most abundant species, the ring-tailed lemur *Lemur catta*, tends to live in rocky crevasses in the arid, sparsely wooded regions. The

Frans Lanting/Minden Pictures

Frans Lanting/Minden Pictures

Frans Lanting/Minden Pictures

Frans Lanting/Minden Pictures

Nimble and playful, the ring-tailed mongoose Galidia elegans *lives by its wits, eating almost anything small enough to capture, from eggs to lemurs. A local legend warns that to laugh at a mongooses's antics is to risk being lost forever in the forest.*

Madagascar lemurs have been studied extensively by scientists because the evolutionary relationships between species can be investigated in a relatively small area, under the same conditions that may have determined the course of their evolution.

There are 11 species of carnivore, including the ring-tailed mongoose *Galidia elegans*, an opportunistic carnivore with a varied diet, and the cat-like fossa *Cryptoprocta ferox*, a nocturnal animal, which hunts birds and lemurs. Despite its claim as the largest carnivore on Madagascar, the fossa is only about twice the size of a house cat. It lives mainly in the trees, often resting in the forks of branches, its long and heavy tail helping it balance. Fossas communicate with scent, using a strong-smelling oil produced from special glands. Deposits of this oil help them find each other in order to mate during a specific period each year; the rest of the year they are solitary animals.

Madagascar is also home to half of the world's species of chameleons and several primitive insectivores called tenrecs *Tenrec ecaudatus*, which are distantly related to hedgehogs. Some tenrecs have

Frans Lanting/Minden Pictures

The nocturnal, solitary, and secretive aye-aye Daubentonia madagascariensis *eats fruit by gnawing through the outer skin and scooping out the pulp with its long, wiry middle finger.*

spines, others do not, and some behave like water rats. Their closest living relatives occur only in the Congo (the giant water shrew) and in the Caribbean (the solenodon).

The destructive impact of humans on the habitats of Madagascar has inevitably led to dwindling numbers of species. The tree-dwelling aye-aye *Daubentonia madagascariensis*, for example, is now virtually extinct, with only a few remaining in the wild. A group of aye-aye has been established on an island reserve to try to protect the species, but aye-aye are solitary animals, and they give birth to only one offspring at a time. These habits make it difficult to build up the population numbers quickly.

International conservation programs aimed at maintaining the unique fauna and flora have been implemented on Madagascar. Although the destruction of the natural vegetation, and hence the habitats for all these unusual animals, has been largely halted, active management is required in order to conserve the diversity that remains. ■

DIANA WALKER

THE LEMURS OF MADAGASCAR

About 70 million years ago, a raft of floating vegetation possibly carried a pair of monkey-like animals from Africa to the island of Madagascar. Finding no other mammalian competitors, these colonists prospered, and their descendants evolved to form a group of at least 40 distinct species, of which about 22 are (barely) alive today. The details of their evolution are not clear, but current research may help to piece together scattered clues from fossil remains and from patterns of present distribution.

KEY TO ANIMALS

1 Brown lemur *Lemur fulvus albifrons*
2 Sportive lemur *Lepilemur mustelinus*
3 Lesser mouse lemur *Microcebus rufus*
4 Golden bamboo lemur *Hapalemur aureus*
5 Aye-aye *Daubentonia madagascariensis*
6 Crowned lemur *Lemur coronatus*
7 Indri *Indri indri*
8 Verreaux's sifaka *Propithecus verreauxi*
9 Ring-tailed lemur *Lemur catta*
10 Verreaux's sifaka *Propithecus verreauxi*
11 Black lemur *Lemur macaco*

Frans Lanting/Minden Pictures

Madagascar's lemurs show an extraordinary diversity in size, habitat, and behavior. The arboreal indri has a body length of about 60 centimeters (2 feet) and is the largest of all living lemurs. It inhabits rainforest and lives in small groups active by day. The lesser mouse lemur is also arboreal but is at the other end of the scale in terms of size. Its body length is about 13 centimeters (5 inches) and it weighs a mere 85 grams (3 ounces). Perhaps the most unusual member of the lemur family is the solitary aye-aye, a night-time forager with bat-like ears. The ring-tailed lemur and Verreaux's sifaka are troop-dwelling lemurs that are active during the day. The ring-tails are terrestrial whereas the sifakas are arboreal. Despite their diversity, almost all the lemurs are gravely threatened by habitat destruction.

Above. *A fat-tailed dwarf lemur peeps out from its home in the hollow of a tree.*

8 Subantarctic and Arctic Islands

STEPHEN GARNETT, KNOWLES KERRY, AND TERENCE LINDSEY

Between the Antarctic continent and the southernmost shores of Africa, Australia, and South America is a series of eight small and remote islands and island groups. These islands, which lie in the subantarctic zone, are the only land within the vast Southern Ocean. Scattered around the globe between 45°S and 59°S, they are as remote from each other as they are from the continents. Although some areas of the Arctic Ocean almost always remain ice-free, nearly all the Arctic islands are interconnected in winter by an enormous blanket of snow and ice.

A Ring of Islands

The subantarctic islands straddle the Antarctic Convergence, an oceanographic boundary formed from the meeting of the cold Antarctic waters with the warmer waters to the north. Those islands to the south of the convergence (South Sandwich, Bouvet, South Georgia, and Heard islands) are extensively glaciated and predominantly snow covered, while those at the convergence (Kerguélen Islands) or a little to the north (Crozet, Macquarie, and Prince Edward and Marion islands) tend to be free of snow for most of the year. Their position in relation to the Antarctic Convergence is important in determining their climate and hence their terrestrial life-forms. Northern islands are green, their lower slopes covered with tussock grasses and herbaceous plants (but no trees). Southern islands have progressively less land free of snow and ice, and the vegetation is sparse and limited to mosses, lichens, and only a few flowering plants on the most southerly ones.

The rockhopper penguin Eudyptes chrysocome *breeds on most of the subantarctic islands, as well as on Tristan da Cunha and several other islands in the south Atlantic.*

The early history of human contact with the islands was one of exploitation, first for fur seals for their pelts, and then for elephant seals and king and macaroni penguins for their oil. Later visits by scientific expeditions raised wider interest in the islands as sanctuaries for seals and seabirds, and for the study of the flora and fauna that have developed in isolation in the subantarctic. The remarkable abundance of seals and seabirds in the unexploited or recovered state has fascinated all who have visited the islands. The surrounding ocean is able to provide food for very large numbers of these species, but the land to which they must return to breed is limited. The Southern Ocean, which encircles Antarctica and flows eastward past the subantarctic islands, was formed during the final stages of the breakup of the supercontinent Gondwana, when the deep oceanic trenches formed between Australia and Antarctica, and then between South America and what is now the Antarctic Peninsula. Until this time the Atlantic and Indian

D. Parer & E. Parer-Cook/AUSCAPE

oceans were separated from the Pacific Ocean, but with the opening of the Drake Passage in the middle to late Oligocene epoch (about 30 million years ago), the oceans began to mix and circulate around the Antarctic continent and the Antarctic Convergence formed. The geographical position of the Antarctic Convergence is moderately constant in present times, but was probably farther north in geological times; then most of today's subantarctic islands were to its south and were, therefore, subjected to a more severe climate.

The geological history of each island is quite different, even though their placement in a ring around the globe might suggest a common origin. Since their formation, the islands have been subjected to relative changes in altitude as they have risen out of the sea. They have also been influenced by changes in sea level brought about by melting of the polar icecaps, and warming or cooling of the surrounding ocean. Since the last glacial maximum some 18,000 years ago, the sea level has risen between 120 and 140 meters (390 and 460 feet) and the glaciers have retreated, thus exposing more and more land. These processes are responsible for the amount of land available for colonization by plants.

The king penguin Aptenodytes patagonicus, *here disporting amid seaweed at the Crozet Islands, is one of several subantarctic penguin species that are increasing in number.*

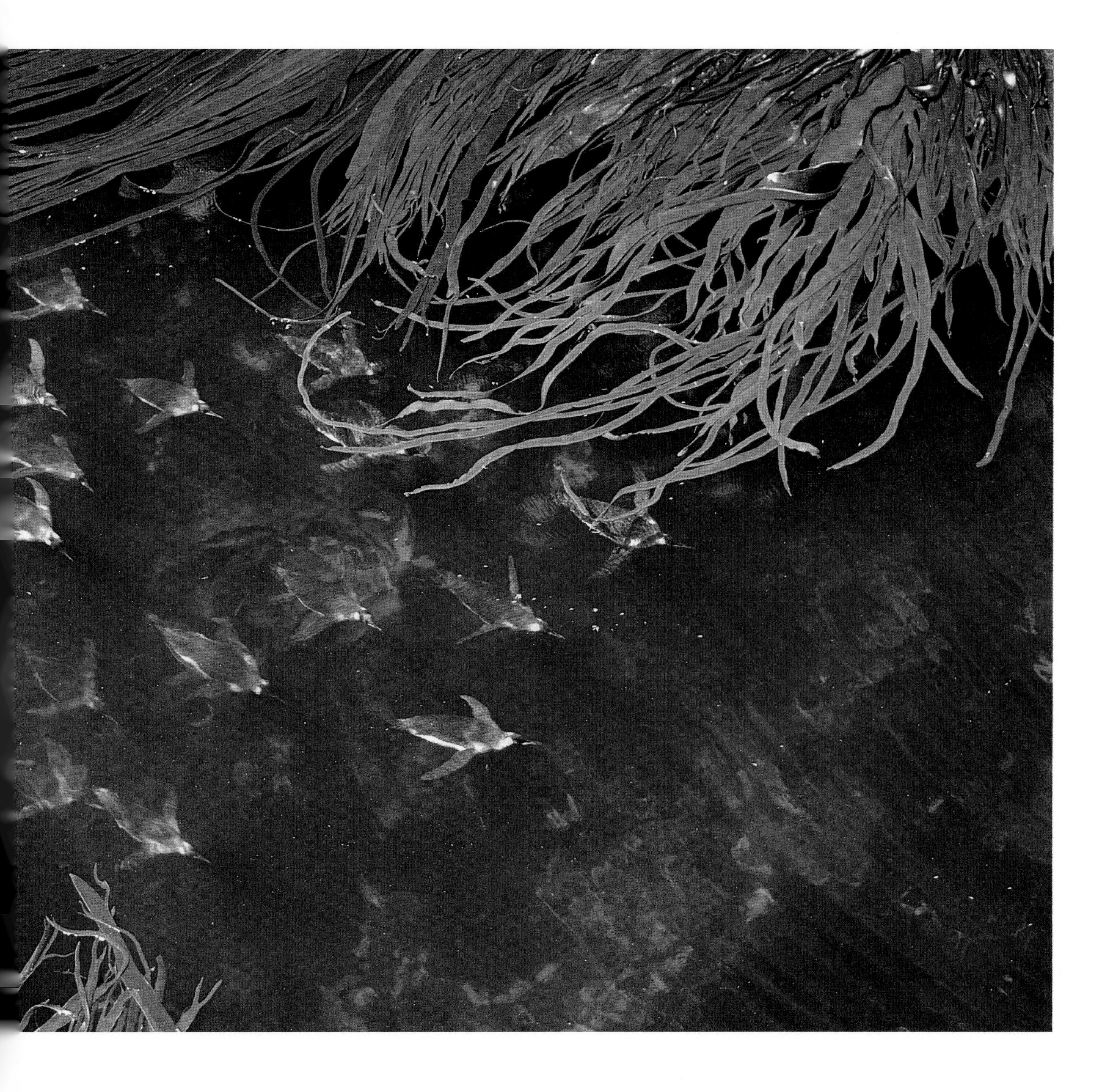

THE DISTRIBUTION OF ISLANDS AROUND ANTARCTICA

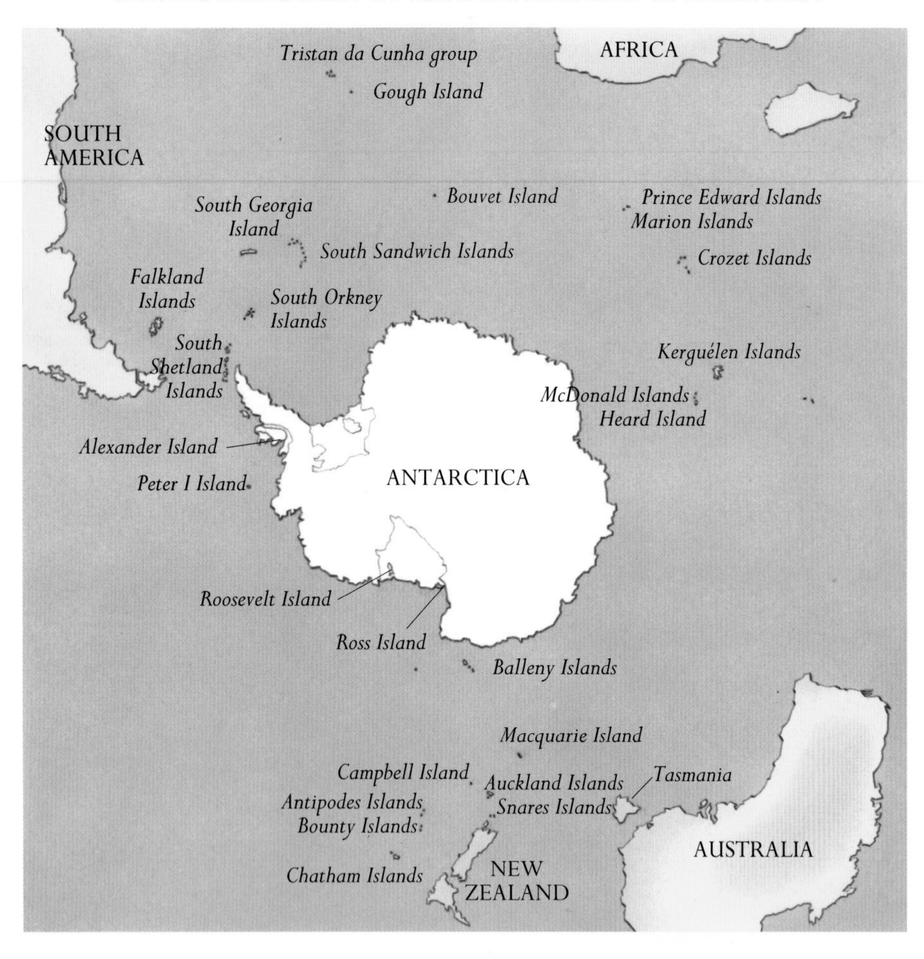

A LIMITED FLORA

The biological history of the islands is interesting and complex, although the range of plants and animals inhabiting any one of them, or in fact all collectively, is small compared with the southern continents to their north. Throughout all the islands, there are only 24 species of grasses, 32 herbaceous plants, and 16 ferns. There are also a minimum of about 250 mosses, 300 lichens, 150 liverworts, and 70 species of mushrooms and toadstools. The few endemic species, and other island flora, show close affinities with their nearest neighbor to the west. Many of the flowering plants are also found on the southern continents; there are, for example, links between South Georgia and Tierra del Fuego; Prince Edward and Marion islands and Africa; and Macquarie Island and Australia.

Jean-Paul Ferrero/AUSCAPE

The world population of about 650,000 southern elephant seals Mirounga leonina *is distributed on subantarctic islands around the Antarctic Convergence.*

The mosses and lichens tend to be cosmopolitan; many are found also in the colder regions of the Northern Hemisphere. The scarcity of subantarctic flora is due to the remoteness of the islands, the direction of the prevailing winds, the area of available land for colonization, and the climate. Only the most northerly islands have any fossil record, and it seems that very little, if any, of the vegetation survived glaciation. The means by which the plants reached the islands is still debated, but it is generally agreed that the present-day flora became established after the end of the last ice age.

Species of flora probably arrived on the islands when their propagules (offshoots) were carried by wind. This is consistent with the similarities between island flora and flora to the west (the prevailing winds circle the globe from west to east at these latitudes). However, it seems likely that other methods of transoceanic dispersal may be involved, including transport by birds with seeds ingested or attached to feathers, and transport by ocean currents. No one method satisfactorily explains the origin of all the plant species.

Survival on land is, of course, as important as survival while being transported over the ocean. The vegetation of the subantarctic islands shows wide tolerance to environmental conditions and in general grows optimally at temperatures higher than those prevailing in the subantarctic. Growth is also limited by lack of sunshine and a marked lack of seasonality in the weather. The species are not necessarily well adapted to the environment, and other species could probably grow better were they to be introduced to the islands.

The vegetation has developed in the absence of vertebrate herbivores and, thus, without grazing pressure. The inability of much of the flora to withstand grazing has been shown where herbivores such as rabbits, sheep, cattle, and horses have been introduced. These have extensively depleted and modified the vegetation, often denuding slopes and leading to soil erosion. This, in turn, has destroyed the habitat of some breeding species and exposed others to the depredations of introduced cats.

ISLAND FAUNA

The truly terrestrial animals of the subantarctic islands are small invertebrate organisms such as nematodes, worms, mollusks (slugs and snails), spiders, and insects, as well as microscopic protozoans, rotifers, and tardigrades. There are no amphibians, reptiles, or fish, and the only vertebrates are seven

Frans Lanting/Minden Pictures

A mother fur seal and her pup. Mammals of the subantarctic region are marine, coming ashore only to breed. The Antarctic fur seal Arctocephalus gazella *has its main breeding colonies on South Georgia.*

species of birds, including sheathbills, ducks, and a pipit. A parakeet and a rail once present on Macquarie Island are now extinct.

The best-known inhabitants of the islands, seabirds and seals, spend most of their time at sea in search of food. They are, however, restricted to land to breed, and during the summer months are present in very large numbers. The most obvious and wide-spread seabirds are the macaroni, king, gentoo, and rockhopper penguins, and the wandering, light-mantled sooty, black-browed, and gray-headed albatrosses. Petrels also breed on the islands in great numbers. These birds, which are related to the albatrosses, range in size: the largest is the giant petrel, which is the size of an albatross, and the smallest are the storm petrels and diving petrels, which are no bigger than sparrows.

Because their adaptations to a life at sea also fit them to survive in the polar regions, and because of the virtual absence of vertebrate predators, seabirds have been able to establish breeding colonies on the subantarctic islands. As breeding sites are limited, the colonies are often very large.

CONSERVING THE ISLANDS

Each subantarctic island is unique in its age and geological history and in its flora and fauna, although some elements are shared. Studies of these islands provide valuable insights into the processes of colonization and the development of communities. Unfortunately, very few islands have escaped the impact of humans, and further, they have been modified by the introduction of organisms, particularly grazing and carnivorous animals, but also species of plants that, having been taken to the islands, end up being able to thrive better than those species that survived ocean crossings. Active programs to eliminate these alien species are now underway on most islands.

Only a few islands are in a near-pristine state. These are important as baselines for the study of all the sub-antarctic islands. There are no introduced species on Heard Island (although seals and penguins were once exploited). The nearby small McDonald Islands may be pristine, as are the South Sandwich and Bouvet islands. The subantarctic islands are wild, rugged, and remote: at once places of beauty and laboratories for science.

KNOWLES KERRY

Frans Lanting/Minden Pictures

Birds of subantarctic regions include those that roam the open ocean except when breeding, such as the sooty albatross Phoebetria, *pictured here in a nesting colony; resident seabirds like the blue-eyed shag* Phalacrocorax atriceps *(opposite top); pirates and predators like the southern skua* Catharacta antarcticus *(opposite center), and egg-stealers like the striated caracara* Phalcoboenus australis *(opposite bottom).*

Birds on Subantarctic Islands

If there is one thing that is always remarked upon by visitors to subantarctic islands, it is the profusion of birdlife. The surge of penguins, the wheeling multitude of petrels, the cacophony of hungry chicks, the stench of regurgitated fish—the ubiquity of the birds is inescapable.

The reason for such a concentration is that just a few islands have to provide safe nesting places for all the birds in the widespread Southern Ocean, an area of abundant food but very little land.

The most prominent members of the subantarctic bird community are the penguins. So abundant are the fish and shrimps around the islands that some colonies exceed a million pairs. Most penguin species are present at their colonies only during the summer, when they come ashore to lay their eggs, raise one or two chicks, then molt. Young king penguins *Aptenodytes patagonicus*, however, grow so slowly that they have to survive all winter on the fat reserves they have built up in summer.

A whole community of smaller seabirds, the petrels, lives inland, among the tussocks and rocks. Each petrel species nests in a slightly different place—in deep burrows, in scrapes beneath tussocks, inside rock crevasses, in slopes of bare scree. They also breed at different times of year, most in summer but some in winter, with the result that there is less competition for food and nesting space.

Petrels are most evident at night. Just after dark the birds that have been feeding on shrimps, fish, or tiny squid at sea return to their nests, relieving their mates that have been brooding eggs or young. They have to be discreet because among the permanent residents of subantarctic islands are several much larger predatory birds, the skuas, giant petrels, kelp gulls, and sheathbills—gull-like birds unique to the islands of the southern Indian and Atlantic oceans. All these birds, which live on spilt food, dead chicks, and seal offal, are quick to take any egg or tiny chick left exposed for more than a few minutes, and are adept at catching tardy prions or storm petrels flying from the island after dawn.

The albatrosses are one group of birds whose size enables them to ignore these predators. Albatrosses are closely related to petrels, both having tubes on their beaks through which they excrete the excess salt they take in with their food, but in general the albatrosses are much larger. The biggest is the wandering albatross *Diomedea exulans*, which has a wingspan that may exceed 3 meters (10 feet). Its long, slender wings are adapted to obtain lift from the slightest breeze, including the gentle movement of the air generated by undulating waves. This large and graceful bird can travel vast distances in search of food.

An odd assortment of bird species is derived from lost migrants that have stayed, bred, and as a result of natural selection over thousands of years, species-radiated into something quite different from their forebears. For instance, most groups of subantarctic islands support a species of cormorant, a fish-eating bird that nests in precarious colonies on cliff ledges. Though all the subantarctic cormorants have derived from one stock, they now differ in facial color and nuances of behavior to such an extent that, even if they encountered each other, they probably would not interbreed. There are also several ducks and shorebirds, a few parrots and rails, and one or two songbirds that have colonized and adapted to their new island homes.

Sadly, even though their islands are so remote, the birds of the subantarctic are not isolated from the ever-growing human influence on our planet. Initially, the effects of colonization were on the islands themselves. Last century, at places such as the Falkland Islands (Malvinas), king penguins were killed in their thousands to extract oil and then devastated by the animals the hunters brought with them. Only now are numbers starting to recover.

More recently, the effects of humans have been felt away from the islands. In particular, longline fishing has decimated albatross numbers over the past 20 years. Albatrosses have learned that boats of any kind are ready sources of food, and poorly thrown baits from the longliners appear to be just another free meal. The hooks in the baits, however, are lethal, killing more than 40,000 albatrosses a year. The population of at least the wandering albatross is in steady decline.

In response to these threats, conservation authorities have had some notable successes removing introduced animals from places like Macquarie Island. They have also developed techniques for preventing birds from taking baits from the longliners, which will benefit both the birds and the fishing parties. In addition, several seabird species, including the chinstrap penguin *Pygoscelis antarctica*, have increased in numbers, probably because there are now so few whales competing with them for krill.

STEPHEN GARNETT

Ben Osborne/Oxford Scientific Films

Frans Lanting/Minden Pictures

Frans Lanting/Minden Pictures

John Shaw/AUSCAPE

Rain-flecked leaves, Pribilof Islands, Alaska. Very low, ground-hugging vegetation is common in the rigorous Arctic environment.

Opposite. *A herd of walrus* Odobenus rosmarus *basking in the sun on the sea ice at one of the New Siberian Islands in the Russian Arctic.*

Arctic Islands

All major landmasses in the north polar region of the world surround an enormous basin, some 14 million square kilometers (5½ million square miles) in extent, that constitutes the Arctic Ocean. Islands lying within the Arctic basin include, in the west, the vast sprawling islands of the Canadian Archipelago (Baffin, Ellesmere, Devon, Banks and Victoria islands, and a host of smaller land fragments), and in the east, Svalbard (Spitzbergen), Franz Joseph Land, Novaja Zemla, North Land, New Siberian Islands, and Wrangel Island (Ostrov Vrangel'a).

Pack ice dominates this environment, effectively removing any distinction between the land and the sea lying beneath it. At its winter height, this blanket of Arctic snow and ice covers some 10 million square kilometers (4 million square miles). As a result, almost all Arctic islands are temporarily interconnected, at least in winter.

The ocean is not always completely frozen over. Even near the pole, shoreleads and, in particular, areas known as polynas remain ice-free almost permanently. One of the largest and most stable of these is the North Water, a region about the size of Switzerland at the northern end of Baffin Bay, between Greenland and southern Ellesmere Island. These "ponds" in a desert of ice play a crucial role in the lives of animals inhabiting the region.

Y. Lanceau/AUSCAPE

Gullfoss Falls, Iceland. An island of contrasts, Iceland has glaciers, active volcanoes, and enough geothermal activity to supply much of the nation's heating and electrical power requirements.

Southward, through the subarctic toward more temperate seas, lie several other groups of islands, including the Aleutian chain, the Commander group (Komandorskije Ostrova), and the Kurils in the northern Pacific; and Iceland, the Faroes, the Orkneys, and several others in the Atlantic.

The southern limit of the Arctic Ocean region is diffuse, merging imperceptibly with waters of the other northern oceans, although researchers believe this may be a geologically recent phenomenon. In any event, the Arctic biome has only recently emerged as a distinct environment, having been remodeled by glaciation during the ice ages.

FLORA AND FAUNA

Few Arctic islands are truly oceanic islands. Though numerous, most lie within a relatively short distance of their nearest mainland. During the ice ages, when vast amounts of water were frozen in the polar icecaps, and sea levels were correspondingly much lower than now, most Arctic islands were joined at some time to the mainland.

In contrast with the tropics, local endemism in either plants or animals is the exception rather than the rule, and in general, Arctic and subarctic island faunas and floras are almost indistinguishable from those of the nearest mainland. Much of the variation that does exist within the region has its origins not so much in the distribution of land and sea (so often the case elsewhere in the world), as in the patterns of refuges established during the past episodes of glaciation. Even at the height of glaciation, some areas remained free of permanent ice, and therefore served as "islands" on which plants (and animals) could survive. These islands spread out as the ice retreated and conditions improved. Much of what is now the Bering Strait region, for example, remained ice-free, and several smaller areas occurred as scattered pockets in the Yukon on the North American continent, and some parts of the Canadian Archipelago.

The Arctic's apparently hostile environment is treeless, but it supports a wide range of plants and animals. Some 1,200 species of plants can be viewed as Arctic in character, although the number falls rapidly northward; northern Ellesmere Island, for example, has only 65 species of flowering plants. Most species are widespread, and extremely localized ones, such as the Spitzbergen buttercup *Ranunculus spitsbergensis* (found only along the shores of a single fiord), are unusual.

Arctic mammals differ little from Arctic plants in their patterns of island distribution. Land-based species like Arctic hares, foxes, and polar bears can reach almost anywhere especially in winter when sea ice connects islands, and when most animals must wander far for food anyway. However, several northern island groups constitute important population centers for pinnipeds (seals and their kin). The Pribilof Islands harbor about 1.3 million northern fur seals *Callorhinus ursinus*, the Commander group about 260,000, and the Kuril Islands about 330,000.

Mark Newman/AUSCAPE

Fur seals and sea lions occur on coasts and islands around the world. This species, Steller's sea lion Eumetopias jubatus, *inhabits the northern Pacific from the Sea of Japan to California.*

The Atlantic puffin Fratercula arctica *is one of the most common Arctic seabirds. It nests in burrows on grassy slopes above coastal cliffs.*

ARCTIC BIRDS

In Arctic regions food resources are abundant during summer, and huge numbers of birds of many species migrate into the area to breed, then retreat to more southerly regions to spend the winter. These include many ducks and geese, gulls, terns and jaegers, as well as a few hawks, owls, and even songbirds. Waders belonging to the families Scolopacidae and Charadriidae swarm over the Arctic in summer, then move southward in some of the most dramatic migrations undertaken by any animal. But northern Arctic islands generally provide no special refuge from mammalian predators (as is the case elsewhere in the world), and most land and water birds breed indiscriminately on island and mainland alike.

It is the seabirds that are most influenced by the distribution of Arctic islands, both for breeding and for winter quarters. In Arctic regions the complex juxtaposition of ocean, islands, sea ice, and continental landmasses means that there are no enormous stretches of strong, steady winds and abundant sea-room so successfully exploited in the Southern Ocean by the albatrosses, shearwaters, and other tube-noses. It is striking that, except for a few storm petrels, only one tube-nose species, the fulmar *Fulmarus glacialis*, is resident in corresponding Arctic latitudes. However, huge numbers of southern shearwaters (notably the sooty shearwater *Puffinus griseus*, short-tailed shearwater *P. tenuirostris*, and Buller's shearwater *P. bulleri*) occupy the northern Pacific and the Bering Sea during the Southern Hemisphere winter, migrating to the Southern Ocean to breed. The resident Arctic marine bird community is dominated by seabirds of other groups—notably ducks, gulls, and auks—all of which are essentially birds of comparatively shallow, near-shore waters, and hence tend to congregate around islands.

Islands below the southern fringes of the Arctic, in north-temperate waters, are almost without exception very small and merely offshore stacks of mainland coasts. However, even these often have an importance to seabird populations out of all proportion to their size and geographical significance. For example, almost all north Atlantic gannets *Morus bassanus* nest on small offshore stacks, sometimes only a kilometer or two from mainland shores. Bass Rock, in the Firth of Forth, Scotland, holds a breeding colony of about 14,000 pairs, and Bonaventure Island in the Gulf of Saint Lawrence, Canada, harbors about 18,000 pairs, together accounting for about 15 percent of the total world population of some 213,000 breeding pairs.

THE AUKS

While the Antarctic and subantarctic regions are dominated by tube-noses, at corresponding latitudes in the Northern Hemisphere another family of birds, the Alcidae, has made the northern Arctic islands conspicuously its own. Auks are small seabirds that feed mainly on small fish caught in shallow waters, and most breed in colonies, large or small. All auks have very high wing-loadings (that is, their wings are small relative to their body weight). Although this arrangement is very effective for flying underwater, it also means that, in the air, they can fly fast but not far. Thus, their powers of dispersal are comparatively limited. All 22 species nest on islands, indeed, mainland breeding sites are small, few, and comparatively insignificant.

Auks range from the High Arctic, where the dovekie or little auk *Alle alle* is sometimes the only species, to temperate regions. Some breed commonly around the British Isles, for example, while a few species extend even farther south—two penetrating as far as the Sea of Cortez, Mexico. One species, the great auk *Pinguinus impennis*, formerly inhabited islands in the northern Atlantic. Largest of its family, it was entirely flightless, and was hunted to extinction for food, the last known pair being killed in Iceland in June 1844.

Some islands in the High Arctic are home to enormous numbers of nesting auks. Prince Leopold Island, for example, is situated in the far north of the Canadian Archipelago in the vicinity of Lancaster Sound, one of the richest marine communities in the Canadian Arctic. This island's dramatic 300-meter (1,000-foot) cliffs serve as breeding sites for about 800,000 seabirds, including about 4,000 pairs of black guillemots *Cepphus grylle*, 86,000 pairs of thick-billed murres *Uria lomvia* (known as Brunnich's guillemot in Europe), as well as multitudes of other birds. Other polar islands have not been so thoroughly surveyed, but little auks breeding in Franz Joseph Land certainly number several million, and another 5 million nest at various colonies in northern Greenland. At least 1 million common murres *Uria aalge* and nearly 2 million thick-billed murres breed on tiny Bear Island, a speck of land just south of Svalbard.

However, auks reach their greatest diversity in the Bering Sea and island chains immediately to the south, including the Aleutian Islands, the Kuril Islands, and the archipelago running south along the Alaska Panhandle. The Pribilof Islands in the Bering Sea constitute perhaps the most impressive example of the intimate link between these birds and their northern island homes: during the breeding season this fog-shrouded island group contains some of the largest concentrations of vertebrate animals to be found anywhere on Earth.

Three species dominate these vast avian cities: the crested auklet *Aethia cristatella*, the least auklet *A. pusilla*, and the tufted puffin *Lunda cirrhata*, although half a dozen other species also nest there in fewer numbers. The birds mass about their colonies in huge swarms, which have been described as "incredible" and "impossible to estimate and difficult to exaggerate". The Aleutians farther south also hold huge numbers of nesting auklets, although the introduction of the blue fox *Alopex lagopus* to sustain local Aleutian economies has had a significant adverse effect on auklet numbers. ■

TERENCE LINDSEY

Frans Lanting/Minden Pictures

Some marine mammals in the Arctic are virtually independent of land, spending their entire lives among the sea ice. One such mammal is the harp seal Phoca groenlandica, *which gives birth to its pups on ice floes.*

John Shaw/AUSCAPE

The crested auklet Aethia cristatella *is perhaps the most abundant seabird in the Bering Sea. It nests in huge numbers on the Pribilof Islands and on Saint Lawrence and Saint Matthew islands.*

R. Powell/Stockshots

PART THREE

Islands and People

All island peoples are colonists.
Using the resources of both land and sea, some colonial settlements thrived and became centers of culture and ideas; others floundered amid the sometimes overwhelming odds of nature and isolation.

9 Two Island Nations

DEBORAH AND PETER ROWLEY-CONWY, AND PAUL MICHAEL TAYLOR

In its short history of 8,000 years of separation from the continent of Europe, Britain has endured wave upon wave of invaders, supplementing or replacing the society that went before it. On the other side of the world, the island nation of Indonesia copes with about 193 million people of varying races, languages, and religions. These two island nations have developed quite differently and yet each has, in its own way, marshaled its resources to become an important world power.

One of Indonesia's chief religious and cultural sites is the huge Buddhist temple complex at Borobudur, Java, which was begun in the late eighth century and completed in the mid-ninth century.

The Indonesian Archipelago

Indonesians refer to their homeland as their *tanah air*, or "land (and) water", for the seas separating the thousands of islands in this archipelagic nation also unite them and form an integral part of the nation. Indonesia straddles the equator between the Philippines to the north and Australia to the south, from the Indian Ocean in the west, eastward to New Guinea. This vast archipelago covers an area of 1,760 kilometers (1,100 miles) from north to south and 5,120 kilometers (3,200 miles) from west to east. Even if the land area only is counted

Michael Freeman/AUSCAPE

Opposite. *Muslim women gather in prayer. Though about 88 percent of the population follows the teachings of Islam, the proportion is not even across the nation: Sumatra is predominantly Muslim, for example, whereas Bali is almost entirely Hindu.*

(2 million square kilometers/772,000 square miles), this constitutes the world's eighth largest nation. With over 193 million people, it is (since the dissolution of the Soviet Union) the world's fourth largest country by population. Cartographers during and after the Dutch colonial era counted 13,677 islands; however satellite images have revealed smaller specks of land and studies suggest that there are more than 17,000 islands within Indonesia's boundaries.

Geographical Groupings

Indonesia's islands are commonly divided into four major groups. The Greater Sunda complex, lying on the sub-oceanic Sunda Shelf, includes the main islands of Sumatra, Java, Borneo (of which Kalimantan is the Indonesian part), Sulawesi, and some smaller islands, such as Bali. The Lesser Sunda complex is formed by the island chain that continues through deep oceans east of Bali, from Lombok to Timor. The Moluccas (or Maluku) form the third region, consisting of deep seas interspersed with volcanic islands and uplifted blocks at the border of the Eurasian and Indo-Australian continental plates. The fourth area is Irian Jaya, on the western (Indonesian) half of the continental island of New Guinea, which sits on the Sahul Shelf and is linked to Australia under the Arafura Sea.

The predominant natural vegetation over the archipelago's landmass is tropical rainforest, typical for a climate of year-round warm temperatures and heavy rain, without a marked dry season. People of almost all Indonesian ethnic groups practice agriculture, although a pre-agricultural form of subsistence is followed by a few hunter-gatherer groups, who sparsely populate heavily forested, interior areas on some outer islands. In the past two centuries population growth and logging have virtually wiped out primary forests on Java, much of Sumatra, and Bali, and the rainforest is now being cut down at an alarming rate on many other islands as well.

Unity in Diversity

Indonesia's seas have both maintained the isolation of its peoples and united them. A long history of inter-island communication has tended to make the coastal populations of all the islands similar to one another, through an exchange of ideas and products that have helped them adapt to the coastal environment. The interior or highland peoples on each island have remained more isolated from each other; their interaction is with the downriver, coastal peoples who maintain contact with other parts of the archipelago. Yet the sea has also prevented mass migrations of people, and along with the difficult-to-traverse interior terrain, it has undoubtedly made it possible for small ethnic groups to survive and, over millennia, develop the tremendous linguistic and cultural

Robert Harding Picture Library

Religion plays a large part in the everyday lives of Indonesians; here a procession makes its way along a country road in Bali.

Robert Harding Picture Library

Indonesia is extraordinary for its ethnic diversity and rich multicultural traditions. Dance is especially important, and every region—in some places every village—has its own troupe, such as this one at Yogyakarta, Java.

diversity that now characterizes the region. Indonesia's national motto, *bhineka tunggal ika,* means "unity in diversity" in Sanskrit. Both these characteristics are sources of this nation's dynamism, tension, and strength. Indonesia is home to over 300 ethnic groups, each with its own language, making this one of the most culturally diverse regions of the world. Yet by the time Europeans arrived in the early sixteenth century, Malay was already being used as a trade language throughout the region. In 1926 advocates of Indonesian independence nominated one widely used Malay dialect as the Indonesian language, and since independence in 1945, its use has been fostered throughout the archipelago.

In the same way that Indonesia's geography has promoted cultural diversity, it has diminished central political control. In a few remote areas of Indonesia's outer islands, tribal groups are still virtually independent of higher-level authorities. This mode of governance was probably predominant throughout the region before the development of the first major kingdoms during the period of Indianized states. During the seventh to fourteenth centuries, the Buddhist kingdom of Srivijaya flourished on Sumatra. At its peak, the Indianized Srivijaya empire reached as far as West Java and the Malay Peninsula. However, by the fourteenth century, the Hindu kingdom of Majapahit had risen in eastern Java, and its leader, Gadjah Mada, succeeded in gaining allegiance from most of what is now modern Indonesia.

Historiography prior to the 1960s explained this period of Indianization by positive civilizing and empire-building migrations, or waves of influence, from India and mainland Southeast Asia, although evidence of actual migrations or conquest is absent. More recently, historians have speculated that indigenous Indonesian chiefdoms independently developed leaders so strong that they actively sought, found, and then imported ideologies, such as Buddhism and Hinduism, to bolster their leadership with notions of divine kingship.

Islam, which had a foothold in Indonesia by the twelfth century, had widely replaced Hinduism on Java and Sumatra by the end of the sixteenth century. Bali and a few other areas, however, actively practice the Hindu religion today—despite some Christian

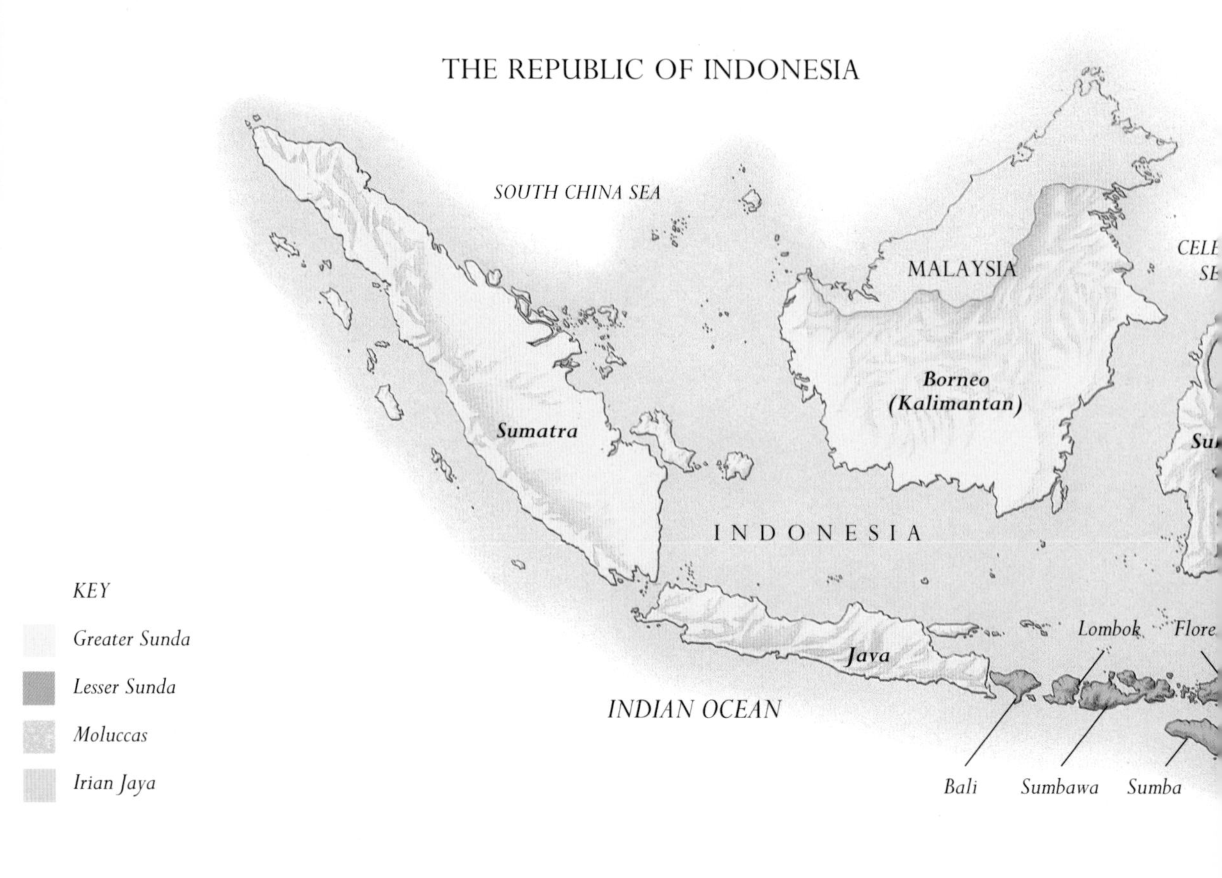

proselytization, which began in the sixteenth century. Currently, about 88 percent of Indonesia's population is Islamic, 9 percent Christian, and 2 percent Hindu.

FROM COLONY TO EMPIRE

As Asia and the New World opened up to Europeans in the sixteenth century, European powers vied for Indonesia's lucrative spice trade. Portugal was the dominant power in this area throughout the seventeenth century, but as early as 1600, the Dutch were slowly establishing themselves as rulers over all present-day Indonesia's islands. The Dutch established their capital at Batavia (now Jakarta) on the main island of Java, and during their 350-year presence, Indonesia developed into one of the world's richest colonial possessions.

The Indonesian independence movement began during the first decade of the twentieth century, expanded in the period between the two world wars, and continued to grow under Japanese occupation during the Second World War. On 17 August 1945, three days after the Japanese surrender, a small group led by Indonesia's later first president, Sukarno, proclaimed independence and established the Republic of Indonesia. Dutch efforts to re-establish complete control met strong resistance, and after four years of warfare and negotiations, the Dutch recognized Indonesia's independence in 1949. The Dutch retained control over the western half of New Guinea, but after further armed clashes and negotiations, full sovereignty over that province was transferred to Indonesia in 1969. East Timor, formerly a Portuguese colony, was annexed in 1976 as Indonesia's twenty-seventh province.

Austral

Sunrise across rice terraces at Tegalalang, Bali. Mechanization is impracticable on these small plots, and about half the total labor force works in agriculture. Rice is the dominant food crop.

Today Indonesia is a republic based on the 1945 constitution, providing for a limited separation of executive, legislative, and judicial power. The president, who appoints the cabinet, is elected for a five-year term. The judiciary, though constitutionally a separate branch of government, is appointed by the executive branch. Legislative authority is divided between the House of Representatives (DPR) and the People's Consultative Assembly (MPR), both renewed every five years. The MPR, which consists of all members of the house plus an equal number of appointed members, meets only once in its five-year term to formulate the overall principles and aims of the government, and to elect the president. Appointed parliamentary members from the military and the government Golkar organization (which serves as the main political party) dominate the house and the MPR.

Indonesia's political system reflects the determination of the government in Jakarta to shift the political focus from Indonesia's deep ethnic and religious differences, which caused the collapse of so many earlier historic attempts to unify the region, via authoritarian, program-based, and development-oriented policies. The country's leaders face such problems as the economy's declining but still considerable overreliance on petroleum, the great income inequality among its people, overpopulation, major regional differences in popular access to the political process, and the incomplete development of civilian institutions independent of the military. Yet this vast archipelago state has attained a healthily growing economy, self-sufficiency in food production, and a steadily rising standard of living—all in sharp contrast to the situation at the time of Indonesia's independence.

PAUL MICHAEL TAYLOR

Alain Compost/Bruce Coleman Ltd.

A worker picking tea in Sumatra. Cash crops such as tea, coffee, and tobacco are important in the central hills of Sumatra and Java, where rice is difficult to grow.

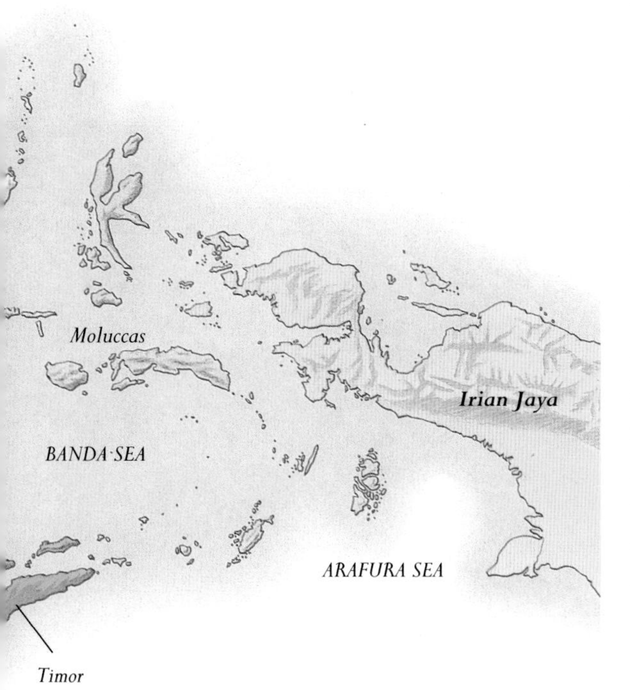

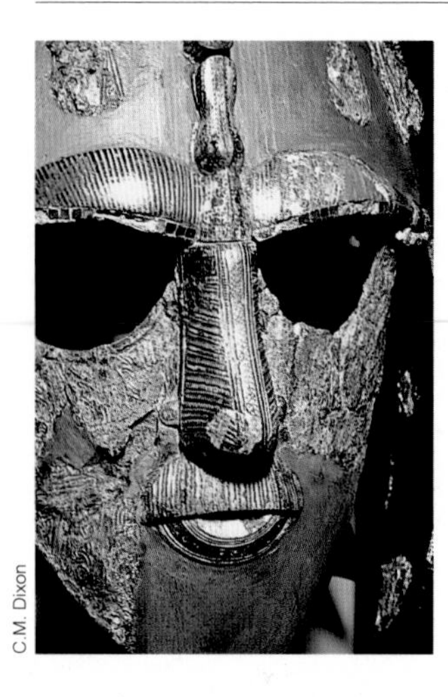

A helmet from a seventh-century Anglo-Saxon burial ship. The Angles, Saxons, and Vikings influenced Britain's development as much as the Romans and the Normans did, but over several hundred years of persistent raids rather than in any single major invasion.

A Fortress Built by Nature?

The British Isles consist of myriad islands dominated by the two largest: Britain and Ireland. Britain, the larger of the two, has remarkable geological and climatic diversity for its size, and this has been enormously important in Britain's history. In sheltered areas on the west coast of Britain, palm trees and other tropical plants can survive out of doors, even in winter, while parts of Scotland would need only a small drop in temperature to initiate the formation of glaciers. The geologically older regions of Scotland, Wales, and northern England are bleak, rugged, and mountainous, while central and southern England are lower lying, with a more temperate climate. For this reason the south has always been more productive agriculturally, and population density has always been greater there—England now has some 50 million people, compared to about five million in Scotland and a little more than three million in Wales. As a result the south and the east of the island have always overshadowed the north and the west.

Britain became an island around 6000 BC, as the glaciers of the last ice age melted and sea levels rose. The last land connection with mainland Europe did not lie between the famous white cliffs of Dover and Calais in France—the intervening lowlands had been flooded some time earlier—but across the present North Sea, via the now-submerged Dogger Bank to the Low Countries (Belgium, Luxembourg, and the Netherlands). The separation had an effect on the culture of Britain almost immediately. Archeologists have demonstrated that new flint arrowhead types, appearing on the mainland shortly after 6000 BC, never spread into Britain.

This medieval manuscript portrays a Viking raid. From the eighth to the eleventh centuries, the Vikings raided and plundered the coasts of Britain.

Farming spread from the mainland into Britain before 3000 BC, but the rectangular houses built by prehistoric farmers on the mainland did not. The inhabitants of Britain continued to build round houses until the Roman conquest shortly after the birth of Christ, which is clear evidence of a separate cultural tradition even though many other cultural influences did cross from the mainland.

The activities of the prehistoric farmers have left a massive legacy, which Britain lives with to this day. Forest clearance was rapid and thorough. Today less than 4 percent of Britain is wooded, and most of the deforestation was carried out before the Roman conquest; Britain 2,000 years ago was, in fact, less forested than Germany is today. Clearance of the thinner woodlands of the northern and western uplands caused the soil to deteriorate and become acidic, resulting in the formation of exposed wet heathlands useless for farming. The deserted heather moorlands, some of the most beautiful and evocative landscapes in Britain, were therefore created not by nature, but by a prehistoric ecological disaster like that now occurring in tropical rainforests.

History reveals two types of conquest of Britain. One was by immigrant groups making piecemeal settlements, forming small political units, and

extending their influence only gradually. The Anglo-Saxon "invasion" after the collapse of Roman power was of this type. Small groups from western Germany infiltrated and settled. Some were employed as mercenaries by the rulers of post-Roman Celtic tribal kingdoms in the south and the east, and from about 430 AD, began taking over these kingdoms and extending their power westward. Two hundred years later, Anglo-Saxon control reached the west coast at Chester, separating the Celtic kingdoms in the area that was to become Wales from those of the north. Warfare was endemic between the various Anglo-Saxon kingdoms, as their leaders sought to extend their territories. The attacks on Britain by Scandinavian Vikings in the ninth and tenth centuries were similar, and much of England and Scotland still bears traces of the success of these raids. Linguistic dialects in eastern Scotland, for example, can still be partly understood by fishermen from western Denmark putting into port to sell their fish, and one famous Scottish clan reveals its Viking origin in its name, Clan Ranald. Some political unity emerged in what is now England, when the surviving Anglo-Saxon kingdom of Wessex reconquered the Viking areas in the tenth and eleventh centuries.

The other type of conquest to which Britain was subject was the landing of a foreign state's army

Unable to defeat the Scots, the Roman emperor Hadrian kept them out of England by building a wall from coast to coast, marking the northern limits of Roman influence. Much of Hadrian's Wall still stands today, like this section at Walltown Crags.

A thousand years after the Romans came, Britain was successfully invaded for the last time when William of Normandy defeated Harold at the Battle of Hastings in 1066, a date known to every English schoolchild.

The Bridgeman Art Library

Pirate or patriot? Regarded by the English as one of their greatest heroes, Sir Francis Drake was a privateer for much of his career, but later commanded the British fleet that destroyed the Spanish Armada in 1588.

Mary Evans Picture Library

intending to subjugate as much of the land as it could. The Roman and Norman conquests were of this type, and although separated by 1,000 years, they have some similarities. The Roman conquest began in 43 AD, and in just four years, the Romans had occupied the south and the east. The bleak and mountainous west (present-day Wales), where the inhabitants resorted to guerrilla warfare, took considerably longer to conquer, while the hostile north was never conquered. In 122 AD the emperor Hadrian began building the wall that still bears his name, stretching 117 kilometers (73 miles) across Britain, from present-day Carlisle on the Solway Firth, to Newcastle on the Tyne estuary. Although heavily garrisoned, Hadrian's Wall was probably just as much a propaganda barrier, designed to daunt the tribespeople to the north and south of it, as it was a practical military barrier.

The Norman conquest followed a similar pattern. The English King Harold, faced with the Norman invasion in the south and the last great Viking raid in the north, was unable to deal with both threats, and was killed in battle by William of Normandy at Hastings in 1066. Again, the southern and eastern lowlands were rapidly subdued. Northern England proved more recalcitrant, and William's Norman army caused massive destruction and depopulation from

1069 to 1070. The frontier with the emergent Kingdom of Scotland was both unclear and troubled. Like the Romans, William and his followers needed to make a statement in stone to cow both his northern subjects and the Scots; but compared to Hadrian their resources were small, so a continuous barrier was out of the question. The outcome was the fortified acropolis of Durham, crowned by the finest Romanesque cathedral in northern Europe, built to the glory of God and the security of the Norman state.

In succeeding centuries the English population and economy grew much faster than those of the north and the west, and so England became increasingly dominant. Wales was subdued by 1300, and when King James VI of Scotland inherited the English throne in 1603, he left for London never to return. Such, by then, was the relative importance of Scotland. Wales and Scotland were a source of soldiers for the armies sent to mainland Europe and elsewhere, and also played a part in the industrial revolution, but in the last few centuries the history of Britain has increasingly become the history of England.

Britain has remained free of invasion from mainland Europe since William landed in 1066, and this has arguably been the most important single influence on its history. Wars have been virtually continuous on the mainland, but Britain has avoided their destructive effects—although frequently sending armies to fight in them. Henry VIII broke with the Roman Catholic Church in the 1530s in common with many other kingdoms of northern Europe, but unlike the others, Britain was not wracked by religious wars. The highly destructive Thirty Years War (1618-48) was fought over almost all continental nations, but the English Civil War (1642–51) was brief and limited by comparison.

The internal political development of Britain, although strongly affected by external events, was therefore not interrupted by foreign armies. The public execution of King Charles I in London in 1649 was an unmistakable symbol of changes in society and economy, and of the growing power of Parliament. When his son was allowed to return in 1660 as King Charles II, he did so as a less than absolute monarch. His successor, James II, showed Roman Catholic inclinations, and was summarily expelled in 1688. Parliament then invited a Dutch prince, William of Orange, to become king. His sole qualification—apart from being married to Mary, a relative of the Stuart dynasty to which James II belonged—was his Protestant religion.

EXPLORER-AUSCAPE

A small section of the 70 meter (230 foot) long Bayeux Tapestry, which depicts scenes from the invasion and conquest of England by William of Normandy.

Mary Evans Picture Library

In a sense, the majestic Durham Cathedral was constructed to serve a purpose similar to that of Hadrian's Wall: as a territorial statement to daunt the Scots.

Political and economic development continued through the eighteenth century despite near-continuous hostilities with France, the dominant mainland military power, from 1688 to 1815.

The fate of France's continental neighbor and opponent Germany shows what might have happened during this period had Britain not been an island: the German principalities were invaded by French armies no fewer than 14 times between 1675 and 1813, an average of once every 10 years.

Unbroken commercial and economic development, coupled with military requirements, were the main reasons for the development of industrialism in Britain during the eighteenth century. Geological diversity provided coal, iron ore, and water power, the raw ingredients for the industrial revolution, very close to

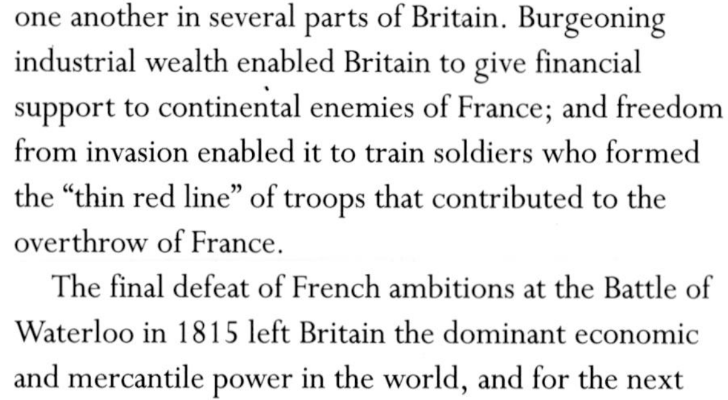

one another in several parts of Britain. Burgeoning industrial wealth enabled Britain to give financial support to continental enemies of France; and freedom from invasion enabled it to train soldiers who formed the "thin red line" of troops that contributed to the overthrow of France.

The final defeat of French ambitions at the Battle of Waterloo in 1815 left Britain the dominant economic and mercantile power in the world, and for the next century the number of its imperial possessions increased to include a large proportion of the world's land. (The twentieth century has seen Britain decline from its dominant position. Some of the other victors of the two world wars—and some losers—have gained economic preeminence.)

Relatively trouble-free political development toward democracy took place, starting with a small extension of the numbers of voters in 1832, and culminating in the granting of votes to women in 1918 and 1928. In 1914 Europe embarked on the First World War. At the time there were major ruling dynasties in Russia, Germany, Austria–Hungary, and Britain; five years later, only the British royal family survived. It is still in place today, although embattled and increasingly isolated compared with the other surviving royal families of Europe, in countries such as Holland, Denmark, and Spain.

It is small wonder, therefore, that insularity continues to be a dominant influence on British consciousness. The phrase "a fortress built by nature" is not a recent description of Britain, but is taken from Shakespeare's play *Richard II*, written in 1595. Shakespeare was writing only a few years after a major invasion scare, namely the sailing of the Spanish Armada in 1588. The fleet from Spain was intended to pick up a Spanish army from the Netherlands and transport it to Britain; but by the time it arrived at the rendezvous the army had gone and the fleet was then destroyed by storm. The contribution of the English fleet to the event was minimal, but sea captains such as Drake, Hawkins, and Raleigh have been elevated to heroic status in British folklore—while the rest of Europe regards them as unprincipled pirates who were ready to prey upon friend and foe alike. Similar near-invasions that are high in the British consciousness are the defeat of the French–Spanish fleet by Nelson at the Battle of Trafalgar in 1805, which prevented a link-up with Napoleon's army of invasion; and the defeat of the German Luftwaffe by the Royal Air Force in the Battle of Britain in 1940, which prevented Hitler's invasion fleet being assembled and launched.

But to envisage Britain as a fortress designed by nature is to overlook the ambivalence of its social

Mary Evans Picture Library

Britain's international power and prestige reached its peak during the reign of Queen Victoria in the late nineteenth century. This Jubilee brochure celebrates her marriage to Prince Albert.

Camera Press Limited/Austral

history. English is only one of the island's three surviving native languages (the others are Welsh and Scots Gaelic, both Celtic in origin. In addition the Cornish and Manx tongues, also Celtic, died out in the 1790s and 1970s respectively, and Norn, a language of Scandinavian origin, was spoken in the Shetland Islands until the nineteenth century). The division between the northwest and the southeast remains important. A still more complex problem is provided by Ireland. Since the Middle Ages, there have been centuries of partial domination of Ireland by Britain. After theFirst World War, Ireland was divided into two parts, one independent and one attached to Britain, but this has not solved the problem; English, Welsh, and Scottish soldiers are still being killed on the streets of Belfast.

Britain's attitude toward the European Community partly reflects its history. The desire of some English people for closer links is tempered in others by an innate suspicion of European institutions, which may take more than a few years of decreased customs regulation and increased exchange controls to dispel. The Welsh and Scottish independence parties are, paradoxically, more enthusiastic about the European Community; their innate suspicion of outside rule is directed in the first instance toward London, not Europe. These parties seek a greater degree of political freedom from England within the context of a more united, federal Europe. Whatever the outcome of the European debate, Britain is likely to continue drinking beer in pints rather than liters for some time to come. ■

DEBORAH AND PETER ROWLEY-CONWY

Tower Bridge and the Tower of London after a night of heavy bombing by the Luftwaffe during the Battle of Britain in 1940.

British soldiers patrol the streets near a recently destroyed building in Northern Ireland.

The Hulton Deutsch Collection/Syndication International

10 ISLAND PEOPLES

BENT FREDSKILD, STUART INDER, KENNETH McPHERSON, AND SIDNEY W. MINTZ

Often envied for their perceived idyllic way of life, island peoples have always had their own uncertainties and turmoil. Vulnerable and isolated, they have needed to become skilled exploiters of the sea for food, trade, and human contact. As well as the basic struggle for survival, most have endured a turbulent history of abandonment, resettlement, plundering, and imperial takeover. Now, many islands have achieved independence and others will inevitably follow.

Jean-Paul Ferrero/AUSCAPE

Climbing for coconuts, Bora Bora, French Polynesia. The copra produced from coconuts is one of the main agricultural exports of Pacific islands.

PACIFIC SEAFARERS AND SETTLERS

The popular view of the Pacific islanders' lifestyle—perpetual ease amid great abundance in a perfect climate—is impossible to withdraw from general currency, because although we recognize it as fiction, it is nevertheless a fiction we would all like to experience. The truth, of course, is that island life is diverse, and that islanders face the same complexity of pressures and problems as communities anywhere.

THE COMING OF THE ISLANDERS

Once, nobody lived in the islands: people found their way to them, over time, by various routes. Just where they came from, and when, and the patterns of their distribution are uncertain, although archeological, linguistic, and other scientific studies are gradually finding answers to these questions. There is wide, if not unanimous, agreement that there was no massive migration, and that the movement into the Pacific islands flowed from the west, through New Guinea, to the east and the north.

At least 40,000 years ago people had found their way to the greater Australian continent, which included Australia and New Guinea, from Asia, by way of Borneo and Indonesia. It took thousands more years before they pushed farther eastward from New Guinea and its associated Melanesian islands into the true insular Pacific. After they had settled Melanesia, others moved into the Micronesian and, finally, the Polynesian islands. There is evidence to suggest that the last wave of expansion, by the ancestors of the Polynesians, was swift, spreading out from New Guinea less than 4,000 years ago.

Great stretches of sea separate the Polynesian and Micronesian islands from Melanesia, and the distribution of people throughout such a huge expanse of ocean was expedited by accidental drift voyages and planned voyages by bold Polynesian navigators in impressively seaworthy craft. Drift voyages are not uncommon today: islanders fishing beyond the reef are sometimes carried away by currents or storms, turning up on islands hundreds, even thousands, of kilometers away many weeks later.

Kevin Deacon/Dive 2000

Although the route east from New Guinea was the major one, undoubtedly some people found their way to the Pacific islands from other directions, particularly from South America (a probability that has been reinforced by several modern raft voyages).

RACIAL GROUPS AND LANGUAGE

Melanesians, Micronesians, and Polynesians are still the main island racial groups, occupying three broad areas of the Pacific. Language, facial features, and skin pigmentation help to distinguish them. The pigmentation of the people of Melanesia (meaning "black islands") is darker than that of the Micronesians ("small islands") and Polynesians ("many islands"). Of the three, the Melanesians generally have the shortest stature and the darkest skin; the Micronesians are taller, their skin lighter; and the Polynesians are the tallest with the lightest skin.

However, for those not particularly concerned with the specialized study of *Homo sapiens*, the division of Pacific island races into Melanesian, Micronesian, and Polynesian is probably no more than a scientific footnote of no great practical relevance. The

Sepik River scene, Papua New Guinea. These people are Melanesians, one of the three broad cultural and racial groups that can be distinguished in the Pacific islands.

C.M. Dixon

A female ancestor figure of carved chalk from New Ireland, Papua New Guinea.

geographical and racial divisions are blurred and there are many exceptions to the rule. Long-established racial pockets are found in islands thousands of kilometers from their "legitimate" groups. Islander characteristics are also being influenced by modern migrations from outside the Pacific, and especially over the past century, by intermarriage.

Between them, the islanders speak more than 1,200 distinct languages; more than 750 are spoken in Papua New Guinea alone. The Pacific's languages belong to either the Austronesian (Malaya–Polynesian) or the Papuan families, most being Austronesian. English is widely spoken in all island groups except the French possessions, where French is the official language. In Melanesia, several versions of pidgin are also used, in some cases as the speaker's first language.

EUROPEAN DISCOVERY AND COLONIZATION

The European discovery of the Pacific came only in comparatively recent times. Not until 1521 did Ferdinand Magellan make his first Pacific landfall, in eastern Polynesia's Tuamotu Archipelago. He sailed on westward, without adding greatly to the world's knowledge of the Pacific islands or their people, for he saw few of either. Gradually, over the next three centuries, others gathered information about the Pacific: a notable amount was recorded by Captain James Cook during his three voyages of exploration between 1769 and 1779.

Having charted the coasts and plotted the islands, from the beginning of the nineteenth century the Western powers began to extend their influence to the people themselves. Missionaries were in the vanguard of the invasion, introducing Christianity in place of the native religions that paid homage to many gods. The first Protestant missionaries, an inexperienced group of well-meaning people brought together in England by the newly formed London Missionary Society, established missions in Tahiti and Tonga in 1797. Christianity is today the major religion of the Pacific islanders.

Missionaries were followed by traders and by government. At the end of the nineteenth century and the beginning of this one, the Western powers divided control of the island groups between themselves, often squabbling or horse-trading over particular islands, but rarely inviting the islanders to participate in the administration and economic exploitation of their lands. The European occupation of the islands was motivated by self-interest: either to control the islands' resources or to hold them as tactical bases against potential enemies and rivals.

The diversity of Pacific peoples is brought together in this eighteenth-century French illustration celebrating the explorations of Captain James Cook and Comte de la Pérouse.

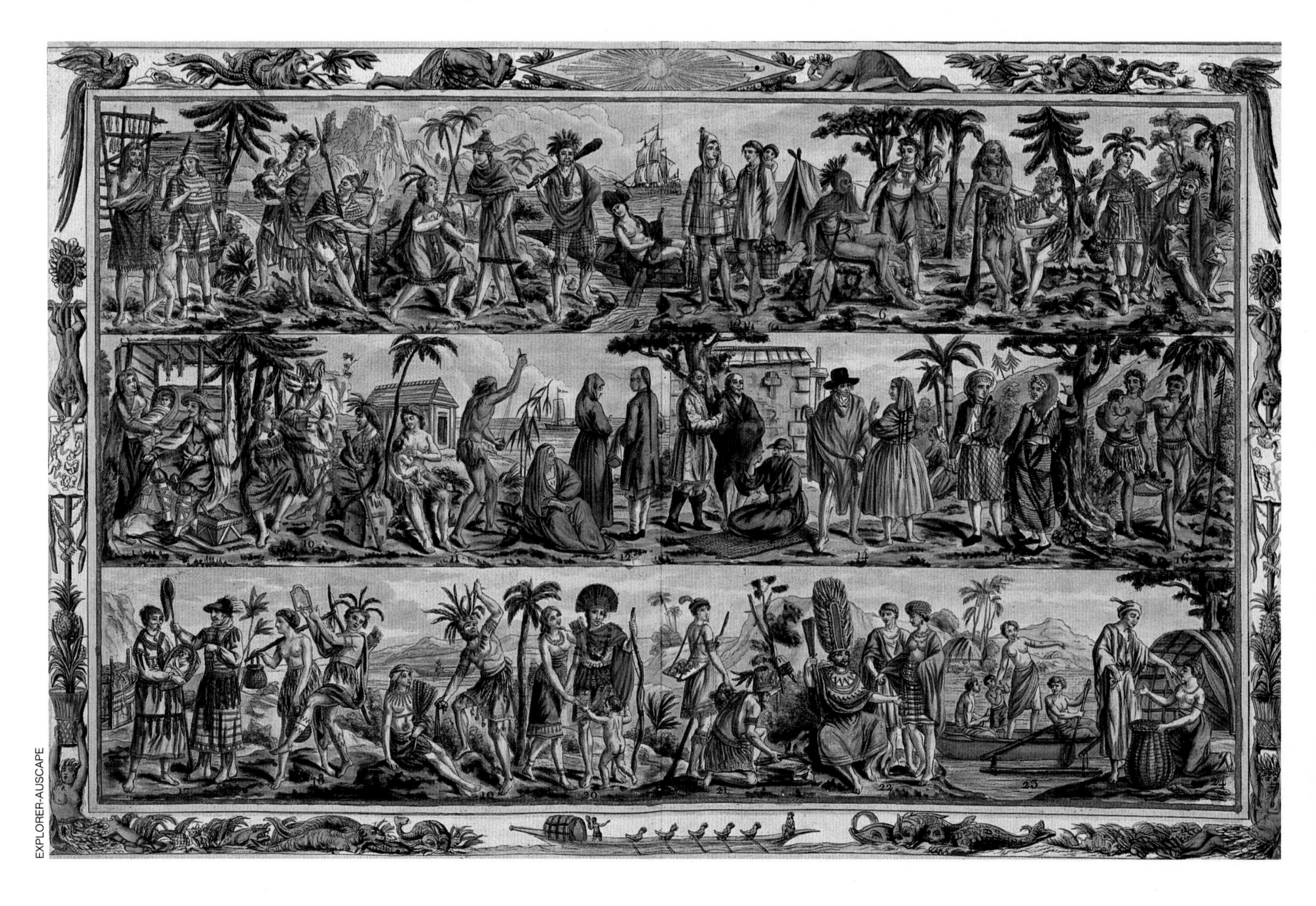

EXPLORER-AUSCAPE

EXPLORER ARCHIVES-AUSCAPE

Pioneer Catholic missionaries in Papua New Guinea. Missionaries preceded traders over much of the Pacific, and in a few remote places still provide the main point of contact with the outside world.

France took possession of Tahiti and New Caledonia, and shared administration of the New Hebrides with Britain. Britain acquired Fiji and its islands, parts of New Guinea, the Solomon Islands, the Gilbert and Ellice islands (including the phosphate-rich Ocean Island), Tonga, and part-control of the New Hebrides. Germany claimed parts of New Guinea, the major Samoan islands, Nauru, and most of the Micronesian islands. The United States took Guam, the Hawaiian Islands, and those Samoan islands Germany had not occupied. New Zealand acquired the Cook Islands and Niue.

PROBLEMS OF COLONIZATION

In the island carve-up, traditional tribal or even racial boundaries were often ignored for the sake of administrative or political convenience. The Gilbert and Ellice islands, for example, were administered by Britain as a single colony although the Gilbert people were Micronesian and the Ellice people Polynesian. The great New Guinea mainland, the western half of which was already claimed by the Dutch, was further divided by drawing a line across the middle: everything south of the line went to Britain, and everything north of the line to Germany. The Samoan Islands were inhabited by one people but split between two powers. Britain and France formalized their joint claim on the New Hebrides with an unusual treaty that gave them both control, yet enabled each to colonize the islands in its own particular way. The arrangement lasted for more than 70 years. Such arbitrary divisions were subsequently to create many problems for the island people themselves.

More seeds of late-flowering dissension were sown when the powers introduced immigrant races: Europeans, Japanese, Chinese, and others from the Asian mainland and Indonesia. Many came as permanent settlers, others as imported laborers for agricultural and mining projects. For 40 years France used New Caledonia as a penal colony for convicts transported from France.

By the time Fiji became independent in 1970, more than half its permanent population was descended from the Indian workers who had been imported by Britain to work the canefields. Concern by the indigenous Fijians that they would lose control to the immigrant population resulted in a coup by the Fijian military in 1987, the overthrowing of the constitution, and restrictions on the political rights of Fiji-Indians.

TOWARD INDEPENDENCE

The end of the First World War brought the first major change in the administration of the islands. Having lost the war, Germany was stripped of its Pacific island possessions which were distributed among Australia, New Zealand, and Japan. More than half of New Guinea and its islands, most of the Micronesian islands, the phosphate island of Nauru, and the Western Samoa group became "mandated" territories, administered by one of the powers on a mandate from, and on behalf of, the League of Nations. In 1933 Japan withdrew from the league and treated its mandated islands as its possessions.

A male ancestor figure of carved and painted chalk from New Ireland.

C.M. Dixon

Austral/Pictor

Jean-Paul Ferrero/AUSCAPE

A New Guinea highlander in ceremonial head-dress. Bright facial paint is part of the traditional body decoration of New Guinea tribespeople and elaborate head-dresses consist of any number of materials including feathers, hair, fur, and leaves.

Above center. *Coconuts, bananas, and other produce laid out for sale, Port Vila marketplace, Vanuatu. Vila is one of the main ports of call for cruise ships touring the South Pacific.*

The end of the Second World War saw further significant changes in island rule, particularly the expulsion of Japanese settlers and administrators. The League of Nations was replaced by the United Nations Organization, and the mandates became "trusteeships", with the United States taking control of the Micronesian islands formerly administered by Japan. Australia made the separate entities of Papua and New Guinea into an administrative union and ruled them as the single territory of Papua New Guinea. France declared Tahiti and New Caledonia to be an integral part of the French republic, and populated New Caledonia with expatriates from Europe and people from other island groups, to the extent that the local Melanesians became a minority.

The trusteeship system, which required the administering powers to make regular reports to the United Nations on conditions in the islands and to admit independent visiting missions on inspection tours, provided a climate in which the islanders could see that their future lay in taking control of their own affairs. This growing political awareness was assisted by a sense of fellowship derived from their membership of the South Pacific Commission, which the powers established in 1947 to enable them to discuss, and perhaps resolve, common economic, technical, and social concerns. Although the Commission did not (and does not) involve itself in political affairs, selected islanders took part in its conferences, and many became leaders in their islands. From widely scattered groups of people, isolated for thousands of years and speaking their own languages, the island populations had come together in the space of only a few years to see that they had common interests and problems.

INDEPENDENCE

The move to independence in the 1960s and 1970s was the most significant change in the Pacific islands since the nineteenth-century scramble for ownership. Decolonization occurred without the conflicts and bloodshed that typified Europe's rush for possession. This comparatively peaceful changeover was as much a tribute to the attitudes and negotiating skills of the islanders, as to the more enlightened policies of the colonial powers.

P. Gauguin 93
aere ia oe -

STRIVING FOR ECONOMIC INDEPENDENCE

Many fragile Pacific island economies benefit today from the millions of much-needed dollars earned yearly from fish caught in their seas. The fishing industry is now a vital source of revenue to governments with few other assets, and which see fishing resources as almost the only hope of improving living standards. Yet only a few years ago the fish in their seas were being plundered by other countries. Furthermore, the catches were being taken away and processed outside the region, the same fish, in cans, eventually finding its way onto the shelves of island stores without the islands having benefited by so much as a cent.

Mark Burgin/AUSCAPE

Traditional net fishing, Malaita, Solomon Islands. Small markets and vast distances restrict development options to little other than fishing and tourism for many Pacific nations seeking economic self-reliance.

The islanders began to exercise significant new bargaining powers over their fishing resources following a decision made by the governments of the South Pacific Forum nations in 1977 to adopt 200 mile Exclusive Economic Zones beyond their coasts. Each island nation was given exclusive rights to the fish and other marine life in its EEZ. The forum then established the South Pacific Forum Fisheries Agency to protect national economic interests in the zones.

The agency soon began using its collective strength to negotiate rights and license fees for foreign fleets to operate in island waters. But the continued insistence of the United States that migratory fish such as tuna belonged to everyone, and that American tuna boats were entitled to follow them into an EEZ without restriction, led to resentment among the Forum nations. In 1984 the Solomon Islands government chased, seized, and confiscated the *Jeanette Diana*, mothership of one of the American tuna fleets, fining its master and owners for poaching, and putting the expensive boat up for sale. This action triggered the United States to impose retaliatory sanctions on Solomon Islands fish exports to the United States, then worth about $US12 million a year, and forming about a sixth of the Solomons' export trade. There followed a diplomatic showdown. With relations between the United States and the Forum island countries threatening to sour, the US State Department intervened, and negotiations resulted in agreement on license fees to be paid to the island governments for all fish caught in their seas.

While this was a significant economic victory, the island nations also derived a new self-assurance from having successfully declined to play by the rules of "big power" politics, when pressing for what they considered to be their rights. It was an important milestone in the islands' efforts to convince the richer powers that they prefer the benefits of trade to aid—that smaller island nations cannot be truly independent without economic independence. A more recent achievement of the Forum Fisheries Agency was the 1991 agreement by Japan to end driftnet fishing in the Pacific. Driftnets, very deep and kilometers long, are walls of death that trap nearly everything in their path.

Fishing, however, is not the islands' only income. Island tourism has also been growing steadily in recent years. For Fiji in 1990 it was the largest foreign exchange earner, a status usually held by that country's sugar industry. Fiji has learned from French Polynesia and Hawaii, both long-established tourist destinations, that a government can encourage growth of the industry through legislation and improvements in infrastructure. Once a workforce and service industries are taught to handle tourism, the tourist industry can attract and absorb millions of investment dollars and expand quickly. And the money is distributed right through the community, which is not true of a mining project, for example, where after the initial development the workforce is likely to be small and specialized, and probably operating remote from centers of population.

Nevertheless, the islands are conscious of the fact that tourism may not always be a firm rock on which to build their economies. It can be sensitive to political and industrial upsets—both local and international—to economic downturns in the areas from which it attracts its visitors and, particularly in the Pacific, to disruption caused by cyclones. And they are aware that they may be forced to face the disruption that tourism can cause to their customs. On that vital point, the Pacific islands are still finding their way. ●

Jean-Paul Ferrero/AUSCAPE

Beachcomber Island, a prominent tourist resort in Fiji. Tourism is the major source of foreign exchange for a number of small Pacific island nations.

Nicholas Devore/Bruce Coleman Ltd.

Friction sometimes arises when traditional fishing vessels, such as these in the Caroline Islands, attempt to compete in their home waters with the far more sophisticated fishing fleets of Japan and other industrialized nations.

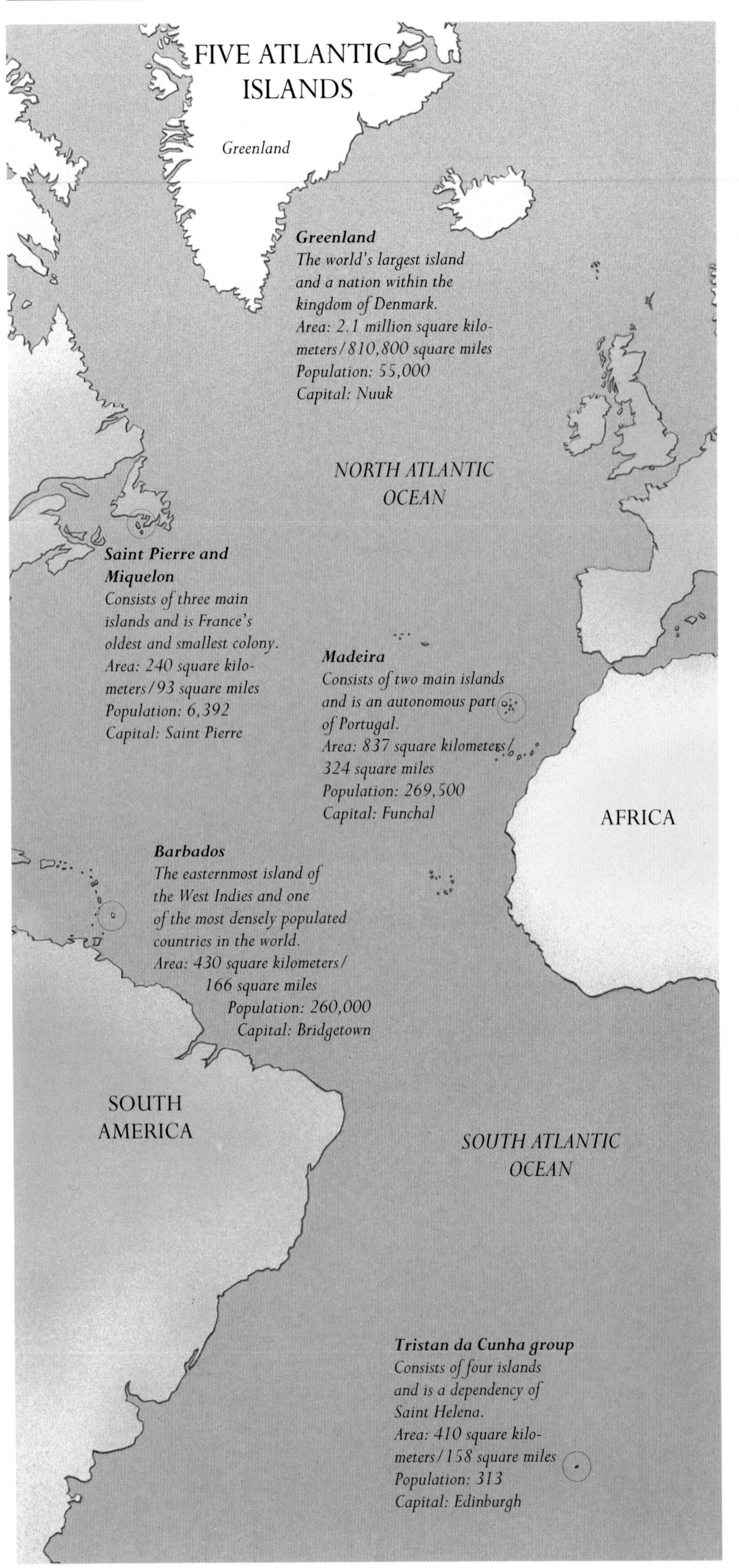

Atlantic Island Life

The Atlantic Ocean did not become part of general European consciousness until the Columbian voyages of the fifteenth century. However, prior to this, from the tenth century on, the Danes had already migrated westward toward the inhospitable shores of Greenland, a large island straddling the Arctic Circle. In fact, the Scandinavian countries, operating quite independently from the rest of western Europe, had ventured as far as North America, via Greenland, long before Columbus.

THE EUROPEAN WORLD VIEW

Before the Columbian voyages, Europe had been accustomed to looking southward and, even more, eastward; its marine center was the Mediterranean Sea, not the Atlantic Ocean. Only after 1492 can we speak of global empires. Before Columbus, "the East" had always lain eastward, not westward; travel, whether overland by the Silk Route or by sea to the eastern Mediterranean, meant eastward travel. The Catalonians, the southern Italians, the Genoese, and the Venetians had all traded and sought to colonize in the eastern Mediterranean. Though many of Europe's scholar-sailors had considered the possibility of sailing westward to reach the East, before Columbus none had attempted it seriously. But political upheavals and disruption of trade routes in the east eventually forced Europe to turn its attentions to the possibilities of the Atlantic.

The measure of this ocean was taken by Europeans only in the course of several centuries, and the widely scattered islands within it were colonized slowly and erratically. While it is impossible to discuss the history of them all, several representative groups are featured.

THREE ISLAND GROUPS

The three island groups of Madeira, Barbados, and Saint Pierre and Miquelon lie far apart and differ greatly. Madeira was an immensely important sugar producer for Europe, long before Columbus set out. Barbados, two centuries after Columbus, was Britain's first "sugar island" and a classic example of British tropical colonialism. Saint Pierre and Miquelon, France's oldest and smallest colony (and its only "white colony"), still maintain the typical north Atlantic way of life.

The Madeira group, believed to be the same lands once identified by Pliny as the "Purple Islands", were uninhabited when discovered (or rediscovered) probably by the Portuguese mariners Vaz and Zarco in 1419. They were claimed by Prince Henry the Navigator, and their settlement was undertaken the following year. Lying about 645 kilometers (400 miles) northeast of Morocco and 965 kilometers (600 miles) southwest of Lisbon, at 32°40'N, these specks of land are at about the same latitude as the island of Bermuda, a British insular possession off the coast of the United States—and about

Nancy Durrell McKenna/The Hutchison Library

Pulled from the water at low tide, these traditional open boats make a colorful addition to the shores of a Madeiran fishing village.

4,830 kilometers (3,000 miles) east of it. Other than several barren rocks, Madeira consists of two islands, Porto Santo (43 square kilometers/17 square miles) and Madeira proper (794 square kilometers/307 square miles). Today they constitute an autonomous region within the Portuguese state.

Barbados is among the most ancient of all overseas British colonies. It is a non-volcanic oceanic island, the easternmost of the Caribbean chain, somewhat isolated from the archipelago, which is stretched north to south from eastern Puerto Rico to Trinidad and the South American mainland. Barbados' nearest neighbor, Saint Vincent, is 160 kilometers (100 miles) to the west. At approximately 13°N 59°W, Barbados is more than 4,845 kilometers (3,000 miles) west and south of Madeira, and nearly 3,508 kilometers (2,175 miles) south and east of Saint Pierre and Miquelon.

Unlike Madeira, uninhabited before Portugal's first settlement in 1420, Barbados had a lengthy pre-Columbian history. Though the Portuguese explorers who discovered it in 1536 found no one living there, archeology indicates that Barbados was populated at least as early as the start of the Christian era, by Amerindian settlers from the South American mainland. Around 1000 AD, Arawakan-speaking migrants also from South America settled there. In 1627, when claimed for the English crown, Barbados was uninhabited, but this tiny island (430 square kilometers/166 square miles) was one of the most densely populated nations in the world by the time it achieved independence from Britain in 1966.

Located at 46°45'N 56°W, Saint Pierre and Miquelon make up France's oldest and smallest colony. There are actually three islands: Saint Pierre, Grande Miquelon, and Langlade–Petite Miquelon, as well as some rocks and cays. The total area of this ancient colony is about 240 square kilometers (93 square miles). It early won a sinister reputation; so dangerous to shipping was it that many called it "the graveyard of the sea". The islands are separated from Newfoundland's south coast by a 24 kilometer (15 mile) channel near the Gulf of Saint Lawrence. As early as 1504, Breton and Norman fishermen seem to have

Mary Evans Picture Library

A contemporary portrait of Christopher Colombus. His discovery of the West Indies in 1492 sparked off a wave of exploitation and settlement that extended across the Atlantic.

Jerry Edmanson/The Photo Library-Sydney

Coconut Beach, Barbados is a popular destination for tourists. Although sugar cane is still the main crop in Barbados, tourism is now the mainstay of the economy.

frequented the islands. But they were not settled by the French until a century later.

Geographically and geologically the three islands are different. Barbados, the vestige of a sunken continent, has a rolling surface built up out of coral limestone. Mount Hillaby, the highest point in Barbados, is only 340 meters (1,115 feet) above sea level. Volcanic in origin, Madeira is strikingly different, with hardly a level surface. The land is described as "crumpled", and its highest point (Pico Ruivo) towers to 1,830 meters (6,000 feet). Different again, Saint Pierre and Miquelon are the remnants of the Appalachian mountains, worn down by glacial movement. Their coast is cliff-lined, their greenery is scant, and there is fog nearly all year round.

THE COLONIAL LEGACY

Not only do these islands differ in location, in terrain, and in climate, but they also represent different eras and different imperial designs. Madeira's history as a sugar-producing Portuguese colony was brilliant but brief. Initiated around 1450, it diminished because of American (especially Brazilian) competition a century later. Slave labor, which would figure so vitally in the New World sugar industry, never acquired much importance in Madeira.

By the mid-sixteenth century, when Barbados had been built upon a sugar economy, the sugar cane in Madeira was being replaced by vines, and Madeira wine began to gain its international fame. Never vital to Portuguese commerce, Madeira has become, in the twentieth century, a picturesque and balmy tourist haven.

Barbados, an important stepping stone in Britain's American empire, was first settled as a farm colony. But in the subsequent two decades it developed—first gradually, then swiftly—into an important sugar-producing colony. As more and more land was bought up to form large estates for producing sugar, the population changed from one of European colonists and smallholders to a few, mostly absentee, plantation owners and masses of enslaved Africans. As sugar mounted in importance in Europe and markets grew, so its production became commonplace in the Caribbean. Barbados soon was eclipsed by Jamaica and felt the competition of other islands. However, from the end of the eighteenth century to the present, the Caribbean sugar industry in general has declined, and Barbados may soon produce no sugar at all. Its economic alternatives are meager: tourism will probably be most important in the future.

To these island societies, remnants of earlier imperialism, Saint Pierre and Miquelon pose a striking contrast. Although some French fishermen had settled there as early as 1670, it was the expulsion of the French from Acadia (Nova Scotia) that gave the colony its first

E. de Malglaive/EXPLORER

Saint Pierre is the administrative and commercial center of the stark, damp islands in the Saint Pierre and Miquelon group. Most of the forests were cleared long ago for fuel and much of the remaining landscape consists of peat bogs.

Rick Strange/The Photo Library-Sydney

Bridgetown, the chief port and tourist center of Barbados, is also one of the most important ports for trade and shipping in the West Indies.

Because of their volcanic origins, the Madeira islands have a dramatic landscape. Many settlements and farms are built on steep slopes and terraces.

stable settlement. Anglo–French rivalry repeatedly disturbed island life; taken and retaken several times in the ensuing half-century, the islands were finally left to the French in 1814, with the understanding that they would not be fortified.

Whereas whale fishing was important in the early years of the colony, cod fishing later emerged as the mainstay of the local economy. In the twentieth century the fishing industry has had to move more and more in the direction of freezing the catch, preparing fishmeal from the leavings, and exporting both products. The larger part of the catch is landed by trawlers and an increasingly minor proportion is taken by small boats.

DISTINCTIVE IDENTITIES

Madeira may be thought of as an island of the first westward expansion because it was part of European awareness before the New World was discovered. Barbados is an island of the second westward expansion, its discovery marked the century when the North European states plunged into the world sugar trade. Saint Pierre and Miquelon were claimed by France only a few years before Barbados was claimed by Britain, yet their history has been strikingly different. So it is that each of these islands has its own story to tell.

SIDNEY W. MINTZ

H. Schmied/ZEFA

Greenland

Greenland is the world's largest island, with an area of slightly more than 2.1 million square kilometers (810,800 square miles), covered for the most part by a giant icecap. About 340,000 square kilometers (130,000 square miles) are ice-free, and these areas are found along the coastal rim, here and there interrupted by huge glaciers in the course of breaking up.

Apart from rocky outcrops, most ice-free areas are covered by dwarf-shrub heaths. However, these coastal strips stretch more than 2,600 kilometers (1,600 miles) from the high Arctic in the north to the subarctic in the south. The polar desert in the extreme north is almost devoid of flowering plants with only two months of the year reaching above 0°C (32°F). In the interior of the south, with seven months above 0°C (32°F), vegetation consists of birch forest and willow copse.

HARDY SETTLERS

Evidence suggests that around 4,500 years ago the first immigrants to northern Greenland came from Canada hunting muskoxen (a woolly oxen) and seals. Fluctuations in climate and prey made living conditions too harsh, and for centuries Greenland seems to have been inhabited only by sporadic waves of these Eskimos.

THE WORLD'S LARGEST ISLAND

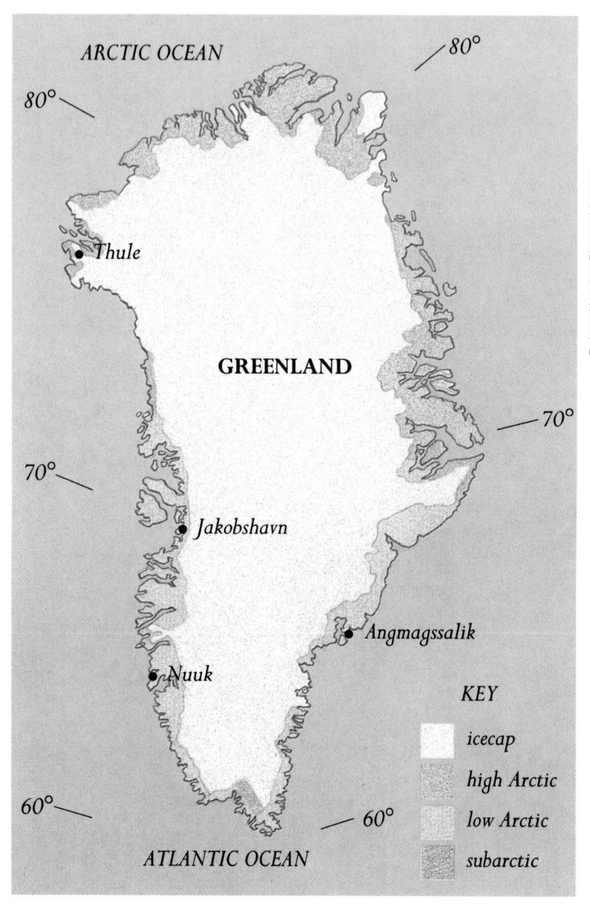

Greenland is covered with an enormous ice sheet second only to that of Antarctica in size. The rigorous climate makes farming all but impossible, and the population of about 55,000 depends on mining, fishing, and fish-processing for income.

Robert Harding Picture Library

In 982 AD Eric the Red was banished from Iceland, which had been colonized a century before by Nordic people. He sailed westward, found the southern Greenland fiords with their luxuriant vegetation, and stayed there with his men for three years. When he returned to Iceland, according to legend, he named the land where he had been Greenland, arguing that settlers would be drawn to go there if the land had an attractive name.

The following centuries saw Norse sheep- and cattle-farmers occupying any arable land in the southern Greenland fiord area and around the head of the fiords at today's capital Nuuk (Godthaab). Farming was supplemented by fishing and sealing. And so Greenland became part of the Danish–Norwegian kingdom. For a

C.M. Dixon

An Eskimo nineteenth-century wooden mask.

The tiny town of Angmagssalik on the east coast of Greenland. An ocean current flowing south from the Arctic Ocean keeps the sea surrounding the town frozen for much of the year. Settlements on the inhospitable east coast are very small and widely scattered.

short period these farmers expanded west to settle in the Labrador–Newfoundland area in North America, several centuries before Columbus. (This small "colony" did not last, possibly because of difficulties with the native inhabitants, or possibly because of being so far from Greenland and trade with Europe.) About the same time as the Norse settlers were moving into southern Greenland, the last wave of Eskimos arrived in northern Greenland from Canada. Of the so-called Thule culture, these Eskimos spread along both coasts to the very south in the following centuries. They are the ancestors of today's native Greenlanders.

A combination of deteriorating climate, overgrazing, and soil erosion led to the extinction of the Norse settlements by the fifteenth century, and for the next two centuries the whale-, seal-, and caribou-hunting Eskimos were the sole inhabitants. During that time, whaling in the north Atlantic became an important industry for the northwestern European countries, and occasionally, the Eskimos bartered their narwhal and walrus tusks with the whalers. In 1721, the Danish-Norwegian kingdom allowed a trading company to send an expedition under the missionary Hans Egede to resettle Greenland—a not unusual combination of religious and commercial colonization.

DANISH ADMINISTRATION

Apart from the introduction of firearms, and stimulants such as tobacco and tea, only minor changes

Fishing boats ride at anchor in the icy waters of Jakobshavn. Fishing, for shrimp, cod, halibut, and salmon, is Greenland's most important industry.

Wally Herbert Collection/Robert Harding Picture Library

occurred in the daily life of the Eskimos with this renewed Danish contact. The Eskimos moved seasonally from one widely dispersed campsite to another, following the availability of game. Gradually, small trading stations with mainly Danish employees grew into towns or "colonies", which attracted some native residents, but official Danish policy was to protect the natives and their traditional way of life.

In the 1950s this policy drastically changed. With the intention of turning Greenland into an industrialized community, its inhabitants working in the fishing industry either on the sea or in fish-processing factories, the Danish government sought to concentrate people in large towns with schools, hospitals, and other public services. Small settlements of 50 people or so, with all the modern amenities, were seen as too expensive. Consequently, most settlements were depopulated by law. At about this time, the inhabitants were given the same legal status as Danes resident in Denmark. By 1 May 1979 Greenland's status was changed to that of a distant nation within the kingdom of Denmark, and home rule was introduced giving Greenland responsibility for its own affairs, except in areas such as foreign and defense policy. Denmark pays block grants to areas such as health service and education, for which self-financing is not possible. Today Greenland's population is 55,000 people, of whom 45,000 are part- or full-blood Eskimo; the 10,000 remaining are Danes. Almost 80 percent live in 17 towns, of which 15 are on the west and south coasts.

Much of Eskimo art draws its inspiration from wildlife and the hunt. These nineteenth-century snow-knives are made from walrus tusk and depict a walrus hunt (top) and harpooning a whale (bottom).

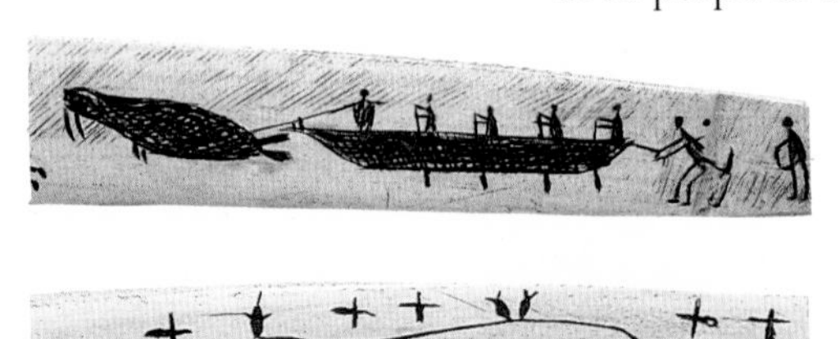

C.M. Dixon

The clean, cold but nutritious waters around Greenland are the habitat of a rich animal life. Cod was once the most important catch, but numbers were drastically reduced as a result of a slight decrease in surface water temperature off western Greenland. Now shrimp is the chief harvest, followed by cod, Greenland halibut, and salmon. Fishing is not the only industry, however. Although Greenland's cryolite, and lead and zinc mines have been exhausted, new finds of gold, platinum, and other minerals promise well for future small-scale mining. Only a few farmers now make their living breeding sheep.

The population of Greenland remains in a vulnerable position, not only because of susceptibility to climate changes and fluctuations in game populations, but also because of the whims of world markets. Decreases in the fishing catch and in the sale price of sealskin have both occurred since the 1980s.

Yet, in spite of everything, a certain optimism characterizes the island. Almost every town has its museum of prehistory and folklore, artists are many, and even a small university teaching in the Inuit (Eskimo) language has been founded. An increasing number of tourists, too, have discovered the grandeur of a sea of drifting icebergs against a green landscape and spectacular plant and animal life—the largest national park in the world.

BENT FREDSKILD

TRISTAN DA CUNHA

SHEENA COUPE

Isolated in the Atlantic Ocean, midway between South Africa and South America and 2,500 kilometers (1,500 miles) from the nearest landmass, Tristan da Cunha is the principal island of a group that contains Inaccessible, Nightingale, and Gough islands. Some 98 square kilometers (38 square miles) in area, the island is almost circular, with an average diameter of 12 kilometers (7 miles). It is dominated by a conical central peak —a legacy of its volcanic origin—whose sides plunge steeply into the sea. The settlement of Edinburgh lies on a plateau at the foot of the cliffs in the northwestern corner.

Although Portuguese navigator Tristão da Cunha sighted the island in 1506, landings were sporadic until 1816, when a British garrison was stationed on Tristan da Cunha to secure nearby Saint Helena where Napoleon was exiled. The garrison was withdrawn a year later, but three settlers remained. With an itinerant population of disaffected seamen, castaways, sealers and whalers, and women from Saint Helena, they founded a new community. They built cottages of stone and thatch, imported livestock, fished the abundant Atlantic waters, and traded fresh vegetables and water with passing ships.

The number of ships calling at Tristan da Cunha fell sharply after 1870 for a number of reasons: whale oil was being replaced by mineral oils, and the American whaling fleet declined during the Civil War; sail gave way to steam on the Atlantic run; and the opening of the Suez Canal in 1869 changed shipping routes. Up to a year might pass between calling ships, and when they left, they usually carried away islanders in search of a more fulfilling life. In 1886 the island supported 97 people. By 1892 the population had dwindled to fifty. The once-prosperous traders resorted to subsistence farming, fishing, and hunting.

Frances Furlong/Survival Anglia

Frances Furlong/Survival Anglia

The settlement of Edinburgh on Tristan da Cunha nestles at the foot of a now dormant volcano. The volcano last erupted in 1961, forcing the temporary evacuation of the entire community to England.
Above right. *Edinburgh, Tristan da Cunha. This traditional thatched cottage reveals an English heritage.*

This subsistence economy continued until the establishment of a naval base in 1942 brought regular radio and shipping contact, and commercial crayfishing began with the opening of a cannery in 1950. Nevertheless, life on Tristan da Cunha remained essentially unchanged until 9 October 1961, when one of the volcanic vents near Edinburgh burst into life. The entire population of 264 was evacuated to England, where the islanders tried unsuccessfully to adjust to modern urban society. Illness and depression, and the disruption of age-old social patterns and hierarchies, convinced almost all to abandon the wider world. An advance party returned to Tristan da Cunha in September 1962, with the rest following in November 1963.

The eruption had left one positive legacy: a tongue of lava had flowed out to sea to form a natural breakwater, which became the foundation of a new harbor. Aid from Britain funded new, more substantial homes and administrative buildings; the pastoral and farming industries were modernized; philately became a significant source of income; a school and hospital were built; and electricity, water, and sewerage systems were installed. Tristan da Cunha today supports a population of 313, still isolated geographically but linked more closely than ever before to the outside world. ●

Adam Woolfitt/Comstock

A Maldive islander brings a basket of coral ashore. A sultanate until the republic was declared in 1968, the Maldives have always been the home of seafarers and traders.

Opposite. *The Miller Atlas (1515–19), a Portuguese nautical chart of the Indian Ocean, reflects the importance of the Indian Ocean as an ancient trade crossroads and meeting place between cultures.*

People of the Indian Ocean

The islands of the Indian Ocean lie on a great curve stretching from Africa to Southeast Asia around the vast emptiness of the ocean. Seaborne human settlement of the islands began at least 40,000 years ago and continued into the nineteenth century, when the last uninhabited islands were settled. Since earliest times, the history and lives of island people have been shaped by the unique monsoonal wind systems of the Indian Ocean and the development of maritime trade.

SETTLING THE ISLANDS

Between 40,000 and 50,000 years ago, in what may have been one of the first seaborne migrations, hunter-gatherers sailed across narrow seas between India and Sri Lanka, and in Southeast Asia, to begin the settlement of Sri Lanka, Australia, New Guinea, and the Philippines. Similar migrations occurred in the following millennia, as seafarers unraveled the secrets of the monsoonal wind systems, and were thus increasingly able to use the ocean for transport and trade.

At least 10,000 years ago, skilled seafarers moved out of southern China into the Malay Peninsula, the Indonesian and Philippine archipelagos, and beyond them east to the islands of Melanesia and Polynesia in the Pacific Ocean. About 2,000 years ago some of their descendants moved west to uninhabited Madagascar, via Sri Lanka and the Maldives, both of which by this time had been settled from India, giving these islands their present linguistic heritage. The Indonesian settlers of Madagascar shaped the languages of that island, and introduced rice cultivation to eastern Africa, as well as the banana, breadfruit, and taro that became a central part of the diet. At approximately the same time, Arab fishing communities also established the first settlements on the Persian Gulf island of Bahrain, where there were plentiful supplies of water, fish, and valuable shells and pearls which were used in trading.

Nevertheless, the importance of the ocean to the island settlers varied enormously. On large islands, such as Madagascar, Sri Lanka, Sumatra, Java, and Luzon, the lives of most people, once settled, revolved around rhythms of agriculture, and only small groups of fisherfolk and maritime traders were affected by the ocean. But for the inhabitants of many smaller islands the ocean was an ever-present influence. Throughout much of the Indonesian and Philippine archipelagos, in the Maldives and Lakshadweep, in Bahrain and on Suqutrā, and along the coast of Africa, island people were dependent upon the sea as a source of food and as the only highway linking scattered communities. The monsoonal winds were vital to their existence. They enabled swift and regular passage by sea and under-

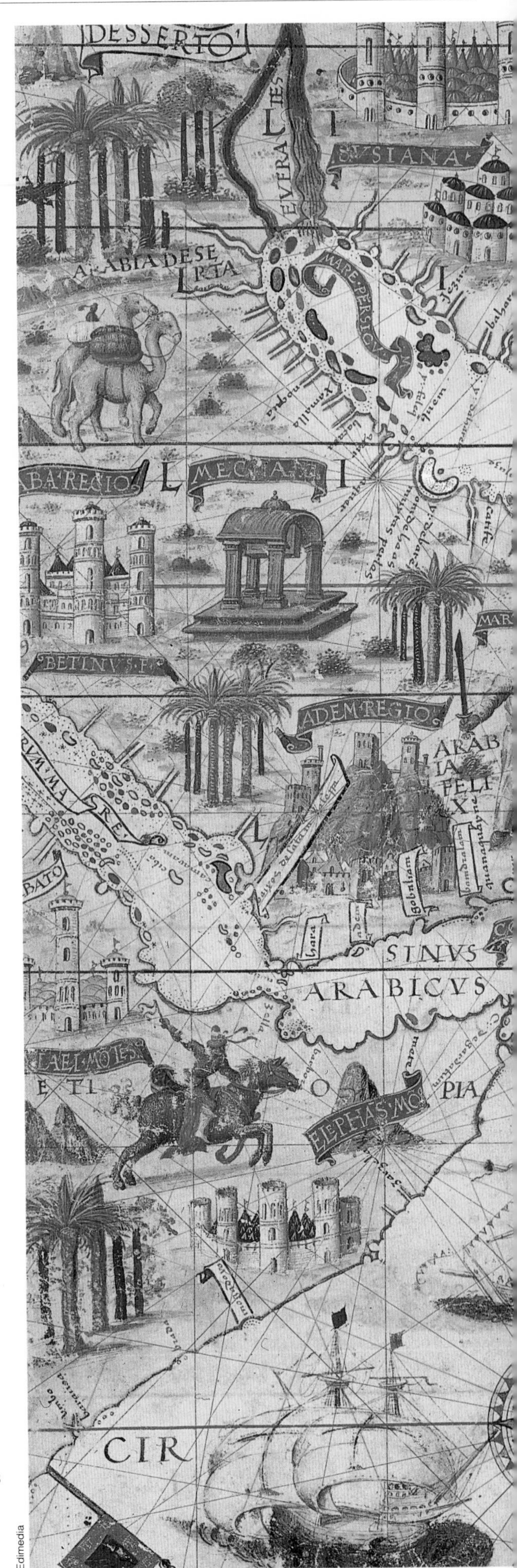

Edimedia

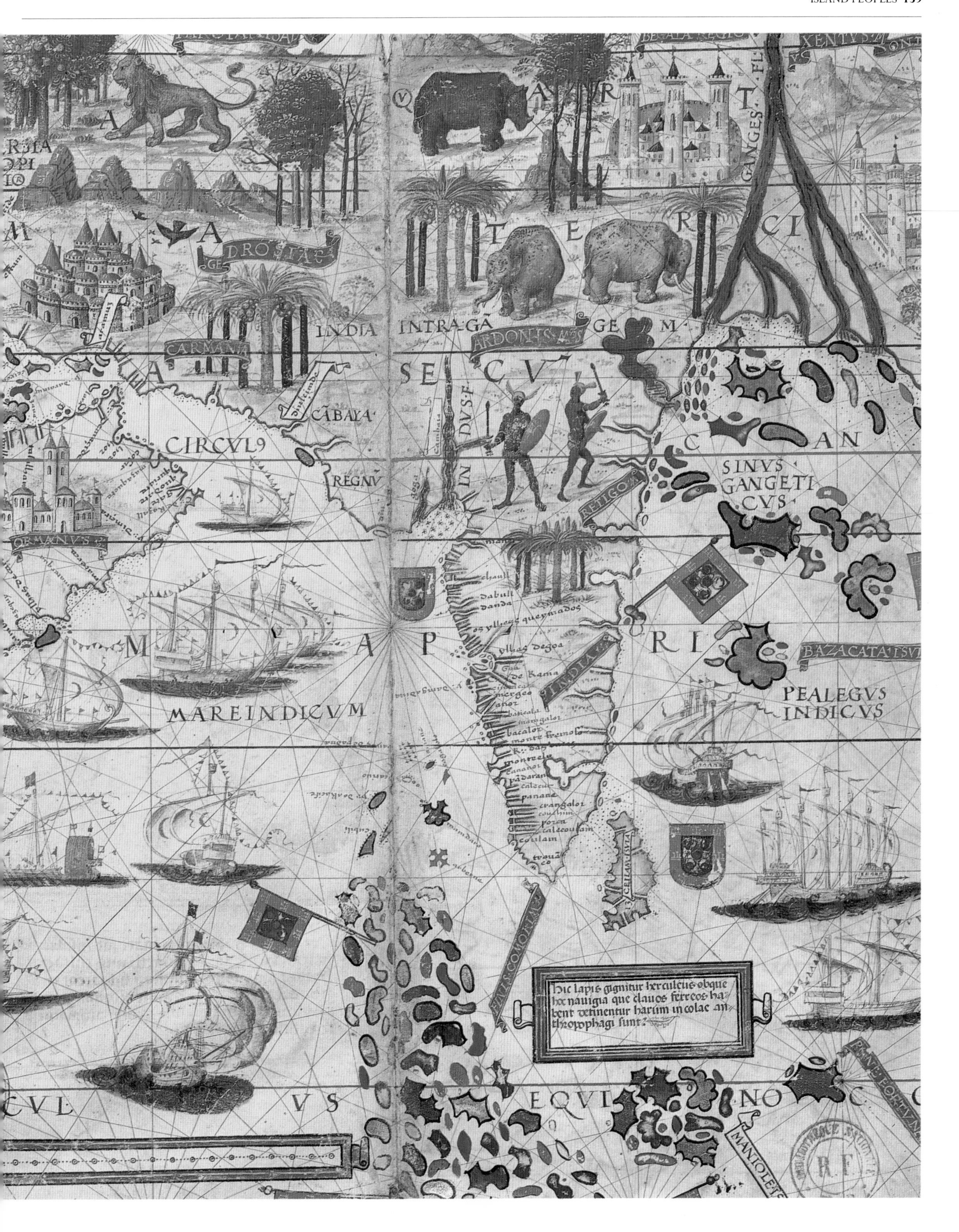
DROSIAE
CARMANA
INDIA
INTRA GĀ
SE
CV
CIRCVL9
CĀBAYA
REGNV
IN DVS F
SINVS
GANGETI
CVS
AN
GANGES
ORMANVS
M
A
P
RI
MAREINDICVM
BAZACATA
PEALEGVS
INDICVS
dabul
danda
os ylheos queymados
ylhas degoa
goa
de Rama
bacalor
monte fremoso
monteely
canonor
cananor
cananor
cranganor
panane
cranganor
coulam
Hic lapis gignitur herculeus obque
hec nauigia que clauos ferreos ha
bent retinentur harum incolae an
thropophagi sunt
CVL
VS
EQVI
NO
C

Coo-ee Historical Picture Library

Raffles Place in Singapore, 1928. In 1819, the British East India administrator Sir Stamford Raffles established the port settlement of Singapore. It rapidly became the center of British colonial activity in the region, achieving self-government in 1959.

Opposite. *Two young Buddhist monks climb a temple staircase. Buddhism spread from India both eastward to Indonesia and as far west as Madagascar, but on many islands has been widely replaced by Islam.*

pinned a complex network of maritime trade which bound the islands into a larger world. Many of these smaller islands were also blessed with valuable resources which attracted foreign sailors and merchants: Suqutrā produced rare resins; Bahrain was famous for its pearls; the islands of eastern Africa were a source of valuable tropical timber, ivory, and ambergris; the Maldives were home to the cowrie shell, used from western Africa to southern China as a form of currency, and to twist products, such as rope and matting found on sailing ships across the Indian Ocean. Many of the smaller Indonesian islands were rich in spices such as cloves and nutmeg, as well as fragrant sandalwood, which was in great demand in China and India.

MARITIME TRADE AND CULTURAL CROSSROADS

During the first millennium AD, with the development of long-distance maritime trade, many of the islands of the Indian Ocean became cosmopolitan cultural crossroads, or were incorporated into a maritime trading network, stretching from Madagascar to the South China Sea and the Philippines. The ports of Sri Lanka, for example, were meeting places for Buddhism and Hinduism from India, and later for Islam from the Middle East; from there Islam spread to the ports of insular Southeast Asia.

Cultural blending also occurred in the Lakshadweep and Maldive islands, which were at the junction of maritime routes linking southern Asia, the Middle East, and eastern Africa. Some 2,000 years ago, these islands were influenced by southern Asian Buddhism and Hinduism. By the thirteenth century, their mercantile links with the Middle East had grown to such an extent that Islam displaced both of these, and a sultanate was established in the Maldives lasting until the 1960s when the islands became a republic. During the same period, many of the offshore African islands, from Suqutrā to the Comoros, converted to Islam and many (for example, Mogadishu, Mombasa, Malindi, Kilwa, Zanzibar and Mozambique) became independent sultanates, whose prosperity depended upon maritime trade and trade with the African interior.

THE EUROPEAN PRESENCE

By the sixteenth century, Muslim seafarers dominated the Indian Ocean, and Arab and Persian words had formed the lingua franca of commerce and navigation. From this time, however, Europeans—Portuguese, Dutch, British, French, and Danish—began to trade in the Indian Ocean, and in the late seventeenth century, the French began the colonization of the uninhabited islands of the Mascarenes (La Réunion, Mauritius, Rodrigues, the Seychelles, and Diego Garcia). These islands were settled both by European colonists and the slaves they imported from Mozambique, Madagascar, the Comoros, and India. These slaves, who formed the majority of the population by the mid-eighteenth century, adopted the religion of their masters and developed their own Afro-French language, Kreol.

Elsewhere during these centuries, the Portuguese and Dutch left racial, linguistic, and cultural legacies in India, Sri Lanka, and throughout the Indonesian archipelago. In India the Portuguese set up enclaves on the western and eastern coasts, where a formidable missionary presence was established. In Sri Lanka they controlled the eastern and southern districts of the island, but were never able to conquer the mountain kingdom of Kandy, which remained a bastion of Buddhist culture. In the Indonesian archipelago, the Dutch established trading posts at ancient ports and colonized the major rice- and spice-producing areas of insular Southeast Asia.

On the eastern limits of the Indonesian archipelago, at the edge of the Pacific, Portuguese and Dutch intrusion was matched by the Spanish, who began their occupation of the Philippines in the sixteenth century. They controlled the islands until ousted by the United States in 1895, leaving behind Roman Catholicism and a culture permeated with Spanish influences. In the southern Philippines, however, the tide of Islam from the Middle East and southern Asia, which had swept across the Indonesian archipelago in the sixteenth century, held firm in the Sulu Archipelago and on Mindanao.

However, the extent of the European legacy should not be exaggerated. Europeans were few in number, their languages did not displace local ones, although they enriched them, and only in some islands and coastal areas did they add to the existing religious diversity. The greatest changes during this period were felt in Sri Lanka and places such as Goa and Kerala off the west coast of India, where Christianity and Portuguese words, names, and music permeated coastal society, and in the Philippines.

In coastal East Africa where they loosely dominated during the sixteenth and seventeenth centuries, Europeans had virtually no impact on Swahili culture and Islam. Madagascar remained in splendid isolation from European contacts until the nineteenth century, when the French took control of the island.

BRITISH INFLUENCE

During the late eighteenth and early nineteenth centuries, the British became the dominant political and economic power in the Indian Ocean region. They conquered most of the French islands, Dutch Sri Lanka, Cape Town and Melaka, and much of the Indian subcontinent. The French were left with La Réunion, and the Dutch with the islands of Indonesia. But the full impact of British domination was not felt until the Industrial Revolution of the nineteenth century, when the lands of the Indian Ocean became the source of huge amounts of foodstuffs and raw materials for the peoples and factories of Europe.

David Beatty/Comstock

C.M. Dixon

Adam Woolfitt/Comstock

Top. *A tea picker at work on a plantation near Kandy, Sri Lanka. The British established coffee plantations in Sri Lanka's central highlands but shifted to tea when disease destroyed the crop. Tea is now one of Sri Lanka's major exports.*

Bottom. *Occupied for centuries, the islands of the Indian Ocean present a fascinating blend of ancient and modern cultures and European, Arab, and Asian influences. Oil-rich Bahrain enjoys a predominantly Arab culture and way of life.*

After slavery was abolished in the British Empire in 1833, plantations across the Indian Ocean were worked by cheap Chinese and Indian labor. (On islands such as Zanzibar and Bahrain, nevertheless, slavery continued for several more generations, and many islands along the coast of East Africa continued to sell slaves to Arab, American, and Brazilian traders who shipped them to the Middle East and the Americas.) Bonded Chinese and Indians worked on rubber plantations and in tin mines in Malaya, and settled as general laborers and skilled artisans on the island entrepôt of Singapore, ruled by the British. Hindu and Muslim laborers, artisans, and petty traders from southern Asia migrated to the sugar plantations, ricefields, and colonial towns of the Caribbean, South America, Burma, Fiji, Kenya, Uganda, South Africa, Sri Lanka, Malaya, Singapore, Mauritius, and La Réunion. They became the backbone of the tea industry in Sri Lanka, and of the sugar plantations in Mauritius and La Réunion (alongside bonded African laborers who remained *de facto* slaves in French colonies until the 1840s). On uninhabited Christmas Island, Chinese and Malay laborers were imported to work the phosphate mine, while on the uninhabited Cocos Islands, a semi-independent feudal plantation state of Malays was created by a Scots merchant family under the protection of the British.

A MELTING POT OF CULTURES

On many Indian Ocean islands throughout the nineteenth century, these new streams of Asian settlers profoundly altered cultural patterns. In the French-speaking Mascarenes, Hinduism and Islam became the religions of the majority, with Christianity the religion of the French-speaking white and black populations. On Mauritius, French was challenged as the dominant language by the evolution of Bhojpuri, a southern Asian dialect related to the Hindi spoken in northern India. Singapore became a Chinese-dominated society, and in Sri Lanka the Tamil Hindu community grew very rapidly. For at least 2,000 years Tamils have inhabited the Jaffna Peninsula in northern Sri Lanka, but with the development of tea plantations in the central districts of the island during the nineteenth century, there was a massive resettlement of Tamils into previously underpopulated mountain regions.

Nineteenth-century colonial rule imposed further cultural influences upon islands of the Indian Ocean. The French occupation of the Comoros and Madagascar had little impact upon the masses in terms of the language and culture of both islands, but Christianity made considerable inroads into Madagascar and upon the Comorian island of Moroni. In addition, the French gave new values and a new language to elite groups who were then able to communicate with a world beyond the Indian Ocean. The same can be said of many of the islands which the British controlled: Mauritius, the Seychelles, Diego Garcia, Bahrain, Sri Lanka, and Singapore. Here the language of administration, trade, and contact with the outside world was English, adding to the compendium of languages already spoken there. In the Philippines the American occupation of 1895 meant English replaced Spanish as the official language.

Elsewhere in the Indian Ocean, island populations were remarkably unaffected by nineteenth-century developments. The Andamans and Nicobars were partially opened up by the British as penal settlements to serve India; the Lakshadweep, Maldives, and Suqutrā slumbered on as remote Islamic corners of the British Empire; while under British rule, the coastal islands of East Africa continued their age-old role as cultural and economic mediators between Africa and Asia. In the southern reaches of the

ZANZIBAR

FRANK H. TALBOT

Zanzibar, and its northern sister Pemba, are two large and fertile islands, the former only 35 kilometers (22 miles) away from the African coast. Zanzibar is a green and fruitful tropical island, scented by drying cloves and copra, set in clean, clear water, and ringed by coral reefs. Almost constant monsoonal winds sing in its great casuarina trees and rattle coconut fronds in the many plantations. There is some surface fresh water on the island, but the secret treasure of Zanzibar is clean fresh water stored deep in caverns in its limestone base. It was this necessary resource that for centuries helped attract settlers and visiting ships.

Robert Harding Picture Library

For 2,000 years or more, Arab trading dhows sailed south from Persia, Oman, and other Arab states carrying carpets, beads, copper utensils, tiles, cloth, china, and salted fish to trade in Zanzibar. Their crews would wait out the hot and humid two to three months between the monsoons, refitting their great ships after using the 4 meter (14 foot) spring tides to beach them. They would load up with tough mangrove poles for house-building, ivory for carving, and rhino horns for making dagger handles, and use the strong southwest monsoon to sail home.

Zanzibar's first inhabitants were probably Africans from the mainland, but the islands are referred to in a Greek text of 60 AD, and they were probably known to southern Arabian and other traders in earlier times. The first known non-African settlement was made by Persians in 701 AD, and not until the early 1500s did Zanzibar and Pemba come under Portuguese domination which lasted for nearly 200 years. The Portuguese were finally ousted by Arabs from Oman who then held dominion for the next two centuries.

P. Maille/EXPLORER-AUSCAPE

"If you play on the flute in Zanzibar, everyone as far as the lakes dances." So goes an old saying, referring to Zanzibar's ancient political and commercial power in East Africa. Central to this power was the slave trade. From the great slave markets on Zanzibar, the French and the Spanish colonists bought slaves for their colonial plantations in Mauritius and South America. The trade was only finally abolished in 1873, because of British pressure. A more respectable industry was found in the growing and trading of cloves, and plantations spread throughout Zanzibar and Pemba in the 1800s; today the bulk of the world's clove crop comes from the two islands.

Zanzibar, now part of Tanzania, remains a fascinating mix of peoples. Its majority is African, but Shirazi (originally non-Arab Persians), Arab, Indian (both Muslim and Hindu), and many others live on the island. The island capital, Zanzibar, still retains reminders of its Arab rule: the Sultan's Palace, built 200 hundred years ago, Persian baths behind the palace built early last century, and many beautiful mosques. The house that explorer David Livingstone stayed in while preparing for his journeys into Africa still stands, as do many beautiful older houses with elaborately carved and bronze-embossed doors. ●

Mary Evans Picture Library

Once a major center for the infamous slave trade, Zanzibar was the only point of access to the interior of "darkest" Africa.

Top. *A huge, ornate door, characteristic of the architecture in Zanzibar.*

Top right. *A Zanzibar street scene.*

Indian Ocean, European whalers established seasonal settlements on previously deserted islands such as Heard, Macdonald, Amsterdam, and Saint Paul islands, and those within the Prince Edward, Crozet, and Kerguélen archipelagos.

In the twentieth century most of these islands have gained independence from colonial rule. Bahrain, Singapore, Sri Lanka, Mauritius, the Seychelles, the Maldives, Madagascar, Indonesia, the Philippines, and the Comoros—with the exception of Mayotte—have achieved nationhood. The islands of coastal East Africa have been absorbed into larger mainland states; the Lakshadweep, Nicobar, and Andaman islands are part of India. Christmas Island, the Cocos group and Heard and Macdonald islands are now part of Australia; Diego Garcia remains a British colony; Prince Edward Island is part of South Africa; and La Réunion, Crozet Island, Kerguélen, Amsterdam and Saint Paul, and the Comorian island of Moroni have been absorbed politically into metropolitan France.

With the exception of Singapore and oil-rich Bahrain, all the smaller islands have remained economically underdeveloped. For the majority, plantations and fishing remain the major sources of income, with tourism rapidly developing as a major foreign-exchange earner. The scattered islands of the southern Indian Ocean still have only seasonal populations of scientists and fishermen. Plantations, fishing, and tourism are important for the larger islands, but they also have significant agricultural bases, and since independence have generally moved toward more complex economies with expanding industrial sectors.

Throughout the centuries the cultures of all the Indian Ocean islands, as well as the coasts of Asia and Africa, have been profoundly affected by interactions stimulated by maritime trade, migration, and colonial rule. Where contact with continental hinterlands was restricted by mountains, climate, and inhospitable terrain, many coastal people were forced to look outward, earning their livelihood through maritime trade and fishing. In the twentieth century, however, coastal peoples have been absorbed by new nation states whose interests generally have turned inward from the sea toward land-based economic development. This has also occurred and accelerated on larger islands such as Madagascar, Sri Lanka, the Indonesian islands of Sumatra and Java, and Luzon in the Philippines, where the spread of modern industrial and mining activity has undermined the contact of coastal peoples with the sea. Only on the smaller islands, where there is currently little scope for the development of modern industrial economies, does the sea continue to play a central role in the people's lives. ■

KENNETH McPHERSON

J. Rufus/Robert Harding Associates

Opposite. *Displaying the catch at Chilaw, Sri Lanka. For many years Sri Lanka was part of a maritime trading network that brought Buddhism, Hinduism, Islam, and Christianity to the island. Today, it endures the effects of bloody conflict between Hindu Tamils and predominantly Buddhist Sinhalese.*

11 The End of Isolation

DAVID HOPLEY

The attributes of isolation and limited size, which make islands so attractive, are also bringing about many changes. The greater the isolation, the more the island has developed as a closed system, sensitive to even the smallest interference. The small size of many islands means that they may quickly reach their carrying capacity for human occupation, while limited resources may place further strains on their ability to sustain rapidly increasing populations.

These problems are more critical for remote and small islands. Volcanic and coral islands lie in mid-oceanic waters that are usually low in nutrients, and poor in fisheries apart from limited resources provided by local reefs. Volcanic soils, although rich and highly productive, are often given over to monoculture, increasing the island's susceptibility to world economic vicissitudes. The mineral resources of the basalt rocks of these islands are very limited and fossil fuels are nonexistent. Guano deposits, which form the basis of phosphatic fertilizers, produce only short-term economic benefits during the period of exploitation. Copra from coconuts may be the only product islands have to offer, at risk from both the physical environment, particularly hurricanes, and the uncertainties of world markets.

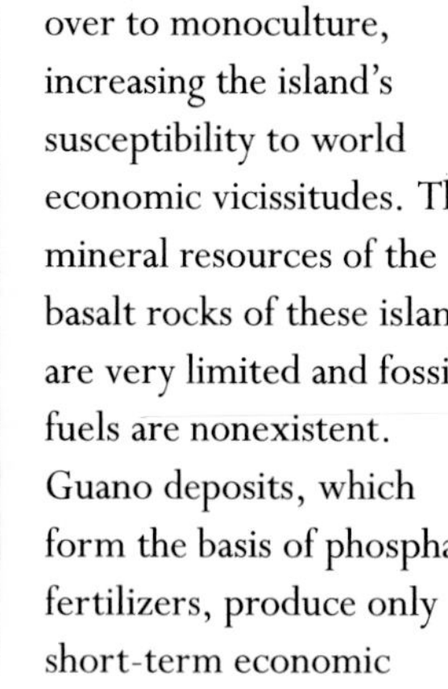

G. Deichmann/AUSCAPE

Copra storage in the Philippines. Prepared from coconut husks, copra is the export mainstay of many tropical island economies.

Opposite. *The grounds of a popular tourist hotel on Oahu, Hawaii. Seen as the ultimate solution for many struggling island economies, tourism may also present problems, such as overcrowding and pollution.*

LIMITED RESOURCES

The uniform nature of many island environments has meant that, even before the advent of modern, economically efficient, large-scale operations, many islands depended on a single product for their living. For high-latitude nations, like Iceland, this has been fishing. The importance of this product, which makes up 75 percent of Iceland's exports, has led to it being central to many of the treaties and trade negotiations Iceland has made with its European neighbors, and ultimately to the naval confrontations known as the Cod Wars of the 1950s and 1970s.

Elsewhere, dependence on single crop agricultural systems typifies narrow island economies. Although the crop may change through time, monoculture is highly susceptible to changes in world markets as well as environmental hazards. In Barbados, and many other islands of the West Indies, the initial staple crop of tobacco, in the sixteenth and seventeenth centuries, was inferior to the product grown on the mainland coast of America. A decline in prices paid for West Indian tobacco led to the introduction of sugar—a labor-intensive crop, which required the import of slaves from Africa. Sugar and its by-products, rum and molasses, subsequently dominated the economy of Barbados, at times providing a rich income, but tied to a colonial system. More recently, however, the sugar industry has declined. Without the capital to mechanize and therefore become cost-effective, the industry now accounts for less than 50 percent of the exports of Barbados.

The island of Salina, only 27 square kilometers (10½ square miles) in area and lying north of Sicily, has traditionally depended on vineyards producing malmsey, a wine once much in demand in England. The vineyards were at their most productive in 1860. By the end of the century, an infestation of the phylloxera parasite destroyed the vineyards. The answer for the islanders was emigration to Argentina and the United States, and later to Australia. In a few decades, the population fell from 8,000 to 2,000 people. Although the vineyards have been re-established, the malmsey market has been wiped out by the popularization of dessert wines such as port, sherry, and others from Portugal and Spain. Emigration has continued. The experience of Salina clearly illustrates the risks involved when a small island bases its economy and survival on a single crop.

CHANGING DEMOGRAPHIC STRUCTURES

Migration has been the response of many islands to worsening economic conditions and overpopulation. The problem of carrying capacity is not new to islands, particularly where subsistence economies operate. In the past, taboos and traditional methods as drastic as infanticide have been applied to control population. The recent trend to migration instead has led to changing demographic and racial balances on many islands, with the potential for developing tensions.

This has occurred, for example, in the Bahamas. Between 1900 and 1920, a large portion of the population left the island to seek work in Florida, cutting

P. Leroux/EXPLORER-AUSCAPE

Nauru was once virtually covered with rich guano deposits, Exported as fertilizer, it gave the Nauru people one of the highest per capita incomes in the world. But now the guano has been exhausted, and the island has no other resources.

cane and laboring in other seasonal harvests. Other workers left to become stevedores in various United States port cities. Then, as tourism started in the Bahamas in the late 1920s, with hotels and infrastructure being built, construction laborers had to be found from other West Indian islands. Since the 1960s, tourism has flourished, and the Bahamas has become a major importer of labor, particularly from the Turks and Caicos Islands, and from Haiti. This, in turn, has produced problems of social tensions, immigration restrictions, and a call for native Bahamians to be given the choice economic opportunities.

Emigration has taken place in many other US trust territories, especially where the opportunity to migrate coincided with the withdrawal of US economic security. In Samoa, the transfer of the island's administration from the US Navy to the Department of the Interior in 1951 resulted in a decline in the defense budget spent on the island, and led to a loss of population and a decline in the economy to subsistence level. There are now as many Samoans living on the west coast of the United States and in Hawaii as are resident in the islands; however, the population in American Samoa has grown over the last 30 years. A concerted economic development effort by the US government led to full employment after the 1960s, and the population rose from 18,000 in 1960 to approximately 156,000 in 1986. However, much of this new increase in population is because of immigration from the neighboring independent island groups of Western Samoa and the Kingdom of Tonga.

Island populations have often been subject to swift demographic changes, from causes such as the seventeenth-century slave trade, which populated the West Indies; or introduced disease, which devastated Pacific islands in the ninteenth century; or massive migrations, such as those discussed. Today, with better health services, fewer opportunities to migrate, and the abandoning of traditional population control practices, many islands will be approaching their maximum capacity in the forseeable future. Signs of this will be uncontrolled urbanization, infrastructure breakdown, environmental degradation, and social collapse. However, increasingly islands are devising more technologies such as waste recycling, and more intensive and productive methods to protect resource renewability.

PROBLEMS FACING SMALL ISLANDS

KEY

● (grey) *Major problem requiring action*

● *Recognized problem*

		PACIFIC ISLANDS	MEDITERRANEAN AND N. ATLANTIC ISLANDS	CARIBBEAN ISLANDS	CANADIAN ISLANDS	CHINESE ISLANDS
RESOURCES	Water resources	● (grey)	●	● (grey)		● (grey)
	Energy	● (grey)	●	● (grey)	●	●
	Overfishing	●	● (grey)		● (grey)	●
	Soil erosion	●	●	●		● (grey)
ECONOMY	Narrow economy incl. monoculture	● (grey)	●	● (grey)	●	
	Population pressure	● (grey)	●	●		●
	Waste disposal	●	● (grey)	●		●
	Tourism	●	●	● (grey)		
MANAGEMENT	Coastal/marine planning	● (grey)	●	● (grey)		● (grey)
	Land tenure	●			● (grey)	
	Wildlife management		● (grey)		●	● (grey)

TOURISM THE PANACEA

Tourism has been hailed as the savior for numerous island economies, particularly in the tropics. There, the climate and environment provide a welcome break for winter visitors from high-latitude, urban environments. The tourist solution is epitomized in the US state of Hawaii, where tourism makes up 35 percent of the gross state product, and the number of visitors exceeds seven million annually. But here, as elsewhere, tourism brings its own problems, with strip development along scenic beaches, alienation of agricultural land—particularly for golf courses—increased demands for limited water supplies, and too high a nutrient content in nearshore waters as a result of such activities as sewage disposal and the application of fertilizers to gardens and golf courses.

Nevertheless, throughout Pacific and Caribbean islands tourism is seen as the most important potential source of foreign exchange to balance the general trade deficits. Development, however, requires foreign investment, which then takes its share of the profits. In the South Pacific it is estimated that only 40 to 50 percent of foreign exchange earnings from tourism remain in the destination country. Up to 30 percent of an island's employment may be tourism-linked, and the problems of a narrowly based economy remain. The tourist industry is particularly sensitive to world affairs, as seen during the Gulf crisis in 1991, which caused an immediate downturn of approximately 20 percent in Hawaii's tourist industry, and is thought responsible for an estimated 10 percent decline in tourists visiting Pacific islands.

Nevertheless, one of the most encouraging aspects of tourism is that it focuses attention on the environment as a resource to be exploited, but only in a sustainable way. The result is that coastal management and planning are being recognized as necessary companions to economic development. This is nothing new to many island societies, whose traditional practices led to conservation. Over-exploitation during periods of European dominance has mainly been responsible for the breakdown of these traditional conservation practices, and today, island nations are realizing that the only way effectively to control pollution and the degradation of fragile environments is to combine traditional laws with modern legislation. This is happening in Pacific island states, led by Papua New Guinea and the Cook Islands. However, the implementation of such planning can tax even the most advanced nations. In 1991 the Hawaii Ocean and Marine Resources Council introduced an Ocean Resources Management Plan, probably the most comprehensive document of its kind in the world. In addition to the obvious problems of conflicting and overuse of limited resources, this sophisticated document listed diffusion of responsibilities, lack of enforcement systems, and reactive, rather than anticipatory, coastal management as major problems.

Greg Vaughn/Tom Stack Associates

Martin Coleman/Planet Earth Pictures

Acute shortages of mineral resources and raw materials bedevil small island economies. This beach on Bali, Indonesia, has suffered severe wave erosion because of the loss of its sheltering fringing reef. The coral was removed to provide lime for cement and other building materials.

Above left. *The development of entirely new industries, such as this ocean "farm" growing giant kelp* Macrocystis pyrifera *on Kona, Hawaii, may profit island economies.*

POLITICAL REORGANIZATION

Smaller island nations are also introducing coastal plans, recognizing that tourism and environmental management go hand in hand. In many instances, this is being carried out as part of regional programs—the most obvious answer to political impotence for small island states, following 100 years or more under colonial rule. While

Michael Jensen/AUSCAPE

The outskirts of Jakarta, Indonesia. In addition to many endemic problems, islands suffer the effects of several much more widespread problems, such as overpopulation and urban sprawl.

ties to a European nation, or the United States, may not have always been to the benefit of the islands, colonialism did give a degree of stability to brittle economies because markets were guaranteed by the political weight of the colonizing government.

For the majority of island states, independence has come in the second half of the twentieth century. Removal of defense infrastructure, and uncertain markets for staples such as sugar, has brought about worsening economic viability. With relatively small populations and small economic resources, individual islands have had little political influence, and little ability to exploit what resources they have. Ratification of the 200 mile Exclusive Economic Zone by the 1982 Law of the Sea Convention means that many island nations potentially have an area of ocean under their jurisdiction several orders of magnitude larger than their land areas. However, without the resources to exploit this area, or to protect it, many islands have turned to political and economic affiliations, a trend that will continue into the twenty-first century.

NO LONGER ISOLATED?

These political affiliations are a recognition that in the modern world islands, however remote, are no longer as geographically or economically isolated as they once were. Jet aircraft link island neighbors, with flights measured in minutes rather than voyages measured in days. Mid-oceanic islands become stepping stones in world communications.

The surrounding ocean is an integrating medium of global proportions, which many islands feel to their cost as well as benefit. Within the first two months of 1993, the wreck of the Liberian-registered supertanker *Braer* shed nearly 600,000 barrels of crude oil into the sea around the Shetland Islands. A few days later the 96,000 tonne *Sanco Honour* collided with the 255,000 tonne *Maersk Navigator*, which was carrying nearly 2 million barrels of oil, spilling it into the Andaman Sea and threatening the coasts of Sumatra and the Nicobar Islands. The world's worst oil spillage took place in 1979, when 2.2 million barrels spilled off the island of Tobago in the Caribbean after two supertankers collided.

P. Leroux/EXPLORER-AUSCAPE

Yves Lanceau/AUSCAPE

Some island nations are concerned about the possible implications of the greenhouse effect. Any rise in sea level would have a disproportionate effect on small cays, such as this one in the Maldives.

Island people have become even more aware of their increasing contact with the outside world because of predictions of global warming from the greenhouse effect. For islands, the important changes are likely to be those associated with a rise in sea level and increased storminess, particularly in tropical latitudes. Although the sea-level rise now predicted is only between 10 centimeters (4 inches) and 40 centimeters (16 inches) by the middle of the next century, the loss of low-lying land is still considered by many to be significant. Particularly vulnerable are the low-lying atoll nations.

In January 1993, the Liberian-registered supertanker Braer *was wrecked off the Shetland Islands, releasing nearly 600,000 barrels of crude oil, resulting in wildlife mortality, fouled beaches, and a large clean-up bill.*

Responses have been as drastic as the suggested removal of whole populations from susceptible islands, or the even more extreme idea that engineering works may be put in place to protect them. However, as pointed out by the Intergovernmental Panel on Climatic Change, the cost is beyond the limited means of the most susceptible islands. For example, it is estimated that it would take up 34.3 percent of the gross national product of the Maldives; more than 10 percent of the gross national product each of Kiribati, Tuvalu, Tokelau, and Anguilla; and more than 5 percent of the gross national product in the case of five other island nations.

While there has clearly been an increase in the greenhouse gases—such as carbon dioxide—in the atmosphere surrounding the Earth, there is still uncertainty about how exactly the atmosphere and ocean will respond. In particular, the part played by the oceans in absorbing carbon dioxide is not fully known, nor is the role of phytoplankton, which live in the ocean. Changes in cloudiness and in distributions of rain, particularly over the highest latitudes of polar regions, may do much to slow down the changes to climate and sea level. The most susceptible coral reef islands may themselves respond by producing more sediment that could, indeed, increase island size with a rise in sea level.

The increasing emergence of confederations of island states will ensure that such global environmental matters will remain high on political agendas. It is generally agreed that whatever the future holds, the return of island ecosystems to as near a pristine condition as is possible will be important, not only in allowing the ecosystems to have the most beneficial response to global change, but for maintaining the sustainability of the most important economic activities of the islands. ■

INTRODUCED SPECIES

TERENCE LINDSEY

The future story of islands and their wildlife may well consist of a battle to control exotic plants and animals, while preventing new invasions, accidental or otherwise. Island ecosystems are generally simpler, and therefore less robust, than mainland systems, and introductions of exotic species of plants and animals are correspondingly more devastating.

ISLAND "LARDERS"

Over the millennia, human settlers of islands have generally taken their food with them, be it goats, pigs, cattle, or other domestic stock. In the days of lengthy explorations by sail in the eighteenth and nineteenth centuries, ships' captains often deliberately stocked remote oceanic islands to serve as "larders"—sometimes with careful planning. When Captain George Vancouver brought cattle to Hawaii in 1794, for example, he also negotiated with the Hawaiian king Kamehameha to impose a ten-year *kapu* (taboo) on their hunting by the islanders, until the animals established themselves.

Geoff Moon/AUSCAPE

The ground-dwelling kakapo builds intricate networks of runways and arenas. It is very close to extinction.

Not all introductions have been of common barnyard animals. In the early 1920s, for example, Norwegian whalers released 11 reindeer on South Georgia. The herd rapidly increased to its current number of several thousand, with drastic effect on the island's fragile natural vegetation.

Of all domestic stock, however, goats inflict perhaps the most severe damage on native vegetation. One of the formidable problems of long-term conservation measures on the Galapagos archipelago is likely to be finding some way of minimizing the effect of goats. The goats, which were introduced as a source of food for settlers, compete with giant tortoises for scarce fodder.

Frans Lanting/Minden Pictures

Introduced to South Georgia in the early 1920s, reindeer have caused widespread destruction of island habitats.

UPSETTING THE BALANCE

Small islands seldom have mammalian predators, and their introduction on many oceanic islands around the world has usually had catastrophic effects. The Indian mongoose *Herpestes auropunctatus* was deliberately introduced into Fiji in 1833, Hawaii in 1883, and Jamaica and various other islands at about the same time. Their purpose was to control rats, but the mongoose includes snakes in its diet, and snakes are prime predators of rats. Mongooses also favor nestling birds. Kauai is the only one of the main Hawaiian islands still free of the mongoose and it seems no coincidence that it is also the island that retains the bird species known to have been there in 1778. A similar situation prevails in Fiji, where Vanua Levu and Viti Levu have mongooses but Taveuni does not.

The story of New Zealand's large, nocturnal and flightless parrot, the kakapo *Strigops habroptilus*, conveys something of the threat that feral cats constitute to island populations. Once common in Fiordland and elsewhere through the early part of the twentieth century, the kakapo declined nearly to the point of extinction. Optimism rose when a population of about 200 was discovered on Stewart Island in 1977. But in studies carried out during the summer of 1981–82, 16 cat-killed kakapos were found, and over the next few years it rapidly became clear that half of the known population was being lost to cats each year. Researchers may even have been partly responsible, their access trails opening up the country. Since it was impractical to clear Stewart Island of feral cats, the entire population of kakapos was captured and relocated to the tiny nearby Codfish Island. Feral cats have also wreaked their havoc elsewhere, notably among the seabirds on Marion and Macquarie islands in the subantarctic.

One of the most dramatic and best-documented cases of decimation by rats began when the trading vessel *Makambo* struck a rock off Lord Howe Island in the Tasman Sea on 14 June 1918. The ship was deliberately beached in an effort to save passengers, crew, and cargo, but rats also left the stricken vessel and disappeared into the forest. Within two years, songbird populations had been devastated and at least four species were extinct. In a recent oral history study, the island's oldest resident was recorded as saying, "People wouldn't believe you now if you told them how many birds there were before the rats got here". Despite half a century of vigorous and sustained extermination efforts, rats remain a serious ecological problem on Lord Howe Island.

Other predators besides mammals have been introduced with damaging effect. In a series of planned introductions in 1949, 1951, and 1952, the barn owl *Tyto alba* was released on the Seychelles in a misguided attempt to control rats. The rats still thrive, but the owls have virtually exterminated the local breeding population of white terns *Gygis alba*.

Udo Hirsch/Bruce Coleman Ltd.

Goats were commonly introduced to islands such as the Galapagos as a source of food for visiting or shipwrecked sailors and early settlers. As herds grew, trampling and grazing by the goats decimated many habitats.

INTRODUCED BIRDS

On islands in many parts of the world, foreign birds have come to dominate the rural environment to such an extent that the casual visitor will see little else but introduced birds, such as starlings, doves, and bulbuls.

Birds have been widely introduced into alien places, either for "aesthetic" reasons, for food, or for sport. Indeed, until the 1920s there existed many effective organizations with the sole intent of fostering such introductions. Semi-wild populations of domestic chickens are now common almost throughout Micronesia and Polynesia. Hawaii has a thriving population of ring-necked pheasants *Phasianus colchicus*, released for sport.

Frans Lanting/Minden Pictures

Left. *Early visitors to the Galapagos Islands introduced horses as a means of transport, but they had little regard for the effect of these animals on the unique flora and fauna of the islands.*

There is, however, an important distinction to be drawn between birds that penetrate native environments and those that merely come to dominate habitats already strongly modified by human activities. Introduced exotics such as starlings *Sturnus vulgaris* and mynahs *Acridotheres tristis* are very common in urban, suburban, and rural areas in Australia, but successful penetration of foreign birds into native habitats such as rainforest or eucalypt woodland has so far been extraordinarily low. This is in marked contrast to New Zealand, where the introduced European chaffinch *Fringilla coelebs* is now one of the most common birds, even in the remote beech forests of Fiordland. Exotic species may compete directly with native species: this may have happened on Norfolk Island, where the island blackbird *Turdus poliocephalus* has disappeared, but its introduced close relative the European blackbird *T. merula* prospers.

THE NEW ARKS

New Zealand's birdlife has been devastated by human activities of various kinds, but it is also favored with a number of small, nearby, off-shore islands. In recent years, wildlife conservation strategies have included the use of some of these islands as kinds of permanently anchored arks, and these have enjoyed considerable success. Several birds that stand almost no chance of survival on the main islands have been transported to islands made free of predators. So far, New Zealand has tended to point the way in this respect: the kakapo on Codfish Island; the stitchbird *Notiomystis cincta* on Little Barrier Island; the little spotted kiwi *Apteryx oweni* on Kapiti Island; and the shore plover *Thinornis novaeseelandiae* on remote Rangatira in the Chatham Islands are only four prominent examples, and more are planned. Such islands need constant monitoring, however. It may be possible to prevent deliberate introductions, but accidental introductions of rats might follow any shipwreck. More insidious still is the threat of introductions of alien plants from seeds lodged in the boots of researchers, or in stores and supplies—possibilities that already worry administrators of Codfish Island, home of the kakapo. Thus it may prove necessary in future to declare some islands completely off limits to any human visitors. •

Ferrero-Labat/AUSCAPE

In 1918, black or ship rats were inadvertently introduced on Lord Howe Island, off the east coast of Australia. They caused a dramatic decrease in local songbird populations.

NOTES ON CONTRIBUTORS

RICHARD S. FISKE
Dr. Fiske is a research geologist at the National Museum of Natural History, Smithsonian Institution, Washington DC. After receiving his PhD in geology in 1960, Dr. Fiske worked at the University of Tokyo, then joined the US Geological Survey, where he carried out research on the active volcanoes of Hawaii, and ancient volcanic rocks of the Sierra Nevada, California. In 1976 he joined the Smithsonian Institution, and served as Director of its National Museum of Natural History from 1980 to 1985. He maintains an active research program that includes the study of explosive submarine volcanoes in Japan, the potentially dangerous volcanoes of the eastern Caribbean, and the mobile south flank of Kilauea Volcano, Hawaii.

PETER G. FLOOD
Dr. Flood is Associate Professor and lecturer in geology at the University of New England, Armidale, Australia, as well as a research associate of the Centre for Sedimentary and Environmental Geology, University of Technology, Queensland. He graduated from the University of New England with first-class honors in geology, and received his PhD from the University of Queensland in 1979. He has since worked on regional mapping programs for the Commonwealth of Australia Bureau of Mineral Resources, in private industry as an exploration geologist, and as a lecturer at the universities of Queensland and New England.

BENT FREDSKILD
Dr. Fredskild is Associate Professor at the Botanical Museum, University of Copenhagen. His areas of study are vegetation history, contemporary flora, and plant geography of Greenland. For almost 20 years he worked in the Department of Natural Sciences at the National Museum of Denmark, participating in excavations of Eskimo and Norse sites in Greenland, and Eskimo sites in Canada. For his PhD, Dr. Fredskild analyzed samples from lakes and bogs for pollen and microscopic plant remains. He has published 100 scientific papers and popular articles, and is leader of the Greenland Botanical Survey.

STEPHEN GARNETT
Dr. Garnett has always been drawn to seabirds. As a child he worked with Don Serventy as he unraveled the story of the short-tailed shearwater in Tasmania, and he has since led or participated in expeditions to study albatrosses, penguins, frigate-birds, and boobies. His research on birds, crocodiles, turtles, and goats has taken him to islands all around the Pacific. More recently, he has been involved in the production of a comprehensive handbook that includes all the seabirds of Australia and Antarctica. Dr. Garnett is a consultant to the Food and Agriculture Organization and the Queensland and Northern Territory governments. He is currently working on the conservation biology of threatened parrots in northern Australia.

RICHARD W. GRIGG
Professor of Oceanography at the University of Hawaii at Manoa, Dr. Grigg is recognized as a world authority on the ecology of both deep-water precious corals and shallow-water reef corals. His research areas include colonization and successional processes of shallow-water coral reefs, and the evolutionary history of fossil reefs in the Hawaiian archipelago. His recent articles on the paleoceanography of coral reefs in the Pacific Ocean published in *Science Magazine* are considered milestones in the field. Dr. Grigg is presently the chairman of the Pacific Science Association Committee on Coral Reefs, and is the managing editor of the journal *Coral Reefs*. He has published over 80 scientific papers, and is the author or co-author of three books.

HAROLD HEATWOLE
Dr. Heatwole is Professor and Head of the Department of Zoology at North Carolina State University, USA. He has taught in universities in Puerto Rico, Australia, and the USA, and has been the president of the Australian Society of Herpetologists, the Great Barrier Reef Committee, and the Australian Coral Reef Society. He is a fellow of the Explorer's Club, the Institute of Biology, and the Australian Institute of Biology,and is the author of more than 200 reports and books on subjects including coral islands, reptiles, eucalyptus dieback, and the Antarctic. He is also on the editorial board of *Fauna of Australia*.

DAVID HOPLEY
Dr. Hopley is Director of the Sir George Fisher Centre for Tropical Marine Studies at James Cook University of North Queensland, Australia. He completed his BA (Hons) and MA at Manchester University, UK, and received his PhD from James Cook University of North Queensland. A geomorphologist, his research for the past 25 years has focused on the evolution of the Australian coastline. In particular, he has examined the effects of changes in sea level on the coastline of Queensland, and studied in detail the evolution of the Great Barrier Reef. His major interests are in the evolution of coral reefs and reef islands and their sustainable management.

STUART INDER
Stuart Inder MBE, is a specialist writer on Pacific affairs. He has been a professional journalist since 1944, and is a contributor to numerous journals, books, and encyclopedias, and a radio commentator on Pacific island matters. He was formerly director of the daily *Fiji Times* Suva, and for nearly 25 years was editor and later publisher of the *Pacific Islands Monthly*, the *Pacific Islands Yearbook*, and many Pacific handbooks, histories, biographies, and guides. He was Pacific affairs writer for the Australian weekly news magazine, the *Bulletin*, from 1982 until 1987. He has been a frequent visitor to all the Pacific island groups and has lived in Japan, Papua New Guinea, Fiji, and Hawaii. Since 1987, he has been a writer with *Australian Geographic*, journal of the Australian Geographic Society.

E. ALISON KAY
Dr. Kay is Professor of Zoology at the University of Hawaii at Manoa and editor-in-chief of *Pacific Science*. Born on Kauai in the Hawaiian Islands, she was educated in Hawaii and California before attending Cambridge University on a Fullbright scholarship. She has served in several administrative positions at the University of Hawaii, including Acting Vice Chancellor, and former chairperson of the Mollusc Specialist Group, Species Survival Commission of the International Union for the Conservation of Nature. She is an honorary associate in malacology of the B.P. Bishop Museum, and President-elect of the American Malacological Union. Dr. Kay is the author of *A Natural History of the Hawaiian Islands: Selected Readings* and *Hawaiian Marine Shells*.

KNOWLES KERRY
Dr. Kerry is a senior research scientist with the Australian Antarctic Division. Since joining the Australian National Antarctic Research Expeditions in 1966 as a biologist, he has wintered on Macquarie Island, and spent many summers there and in Antarctica conducting biological research. As a permanent member of the scientific staff of the Australian Antarctic Division, he established the Australian biological research program in Antarctica in 1972 and the marine biology program in 1979. He served on the Australian delegation which negotiated the Convention for the Conservation of Antarctic Marine Living Resources, and represented Australia at meetings of the scientific committee of the commission. He has recently edited *Antarctic Ecosystems: Ecological Change and Conservation*.

TERENCE LINDSEY
Terence Lindsey was born in England but raised and educated in Canada. Resident in Australia since 1968, he has traveled widely throughout Australasia and the south-west Pacific, and has written several books on birds. He has also acted as editor, contributor, consultant, researcher, and illustrator on many publications. Active in several ornithological societies, he is an associate of the Australian Musuem and a part-time teacher with the Department of Continuing Education, University of Sydney.

KENNETH McPHERSON
Dr. McPherson is Associate Professor and Executive Director of the Indian Ocean Centre for Peace Studies at Curtin University and the University of Western Australia. He established the *Indian Ocean Review* in 1980, the Centre for Indian Ocean Regional Studies in 1987, and the Peace Studies Centre in 1990. He is the author of numerous articles on Indian Ocean history, and is currently working on a major history of the region.

ALEXANDER MALAHOFF
Dr. Malahoff is Professor of Oceanography and Director of the National Oceanic and Atmospheric Administration's (NOAA) National Undersea Research Center (Hawaii Undersea Research Laboratory) at the University of Hawaii at Manoa. His research over the past 25 years has been focused on the geology and geophysics of the ocean floor and of volcanoes and volcanic islands, using ships, airplanes, and submarines in order to conduct these studies. He is the author of over 70 scientific papers; much of his recent work has centered around studies on hydrothermal vent processes and cobalt-rich ferromanganese crusts around the Hawaiian Islands and other areas.

SIDNEY W. MINTZ
Dr. Mintz is William L. Straus Jr. Professor of Anthropology at Johns Hopkins University, Baltimore, USA. He is an anthropologist of the Caribbean region, working on economic anthropology, the history of peasants and proletarians, and the anthropology of food. He has written numerous articles and reviews, and his published works include *The People of Puerto Rico* (co-author), *Worker in the Cane*, *Caribbean Transformations*, *Sweetness and Power*, and most recently, *The Birth of African-American Culture* (co-author).

STORRS L. OLSON
Dr. Olson is Curator of Birds, Department of Vertebrate Zoology, National Museum of Natural History, Smithsonian Institution, Washington DC. His research interests lie in systematics and the evolution of recent and fossil birds of the world, and in human-caused extinctions of birds on oceanic islands. His special focus has been on the islands of Hawaii, the West Indies, and the Atlantic. An authority on fossil birds in general, his first expedition was to Ascension Island in 1970. He has collected fossils on four other islands in the Atlantic, and on several islands in the West Indies. With his wife Helen James, he has also discovered and described dozens of new species of fossil birds in the Hawaiian Islands.

DEBORAH ROWLEY-CONWY
Deborah Rowley-Conwy works for the Durham County Council in the Arts, Libraries and Museums section. After training as a nurse, she changed careers, working for British Airways and British Telecom, and traveling widely. While based in Cambridge, she led historical tours in East Anglia, and later moved north to Durham, where she is now coordinating the events celebrating the 900th anniversary of the founding of Durham Cathedral.

PETER ROWLEY-CONWY
Dr. Rowley-Conwy is a lecturer in archeology at the University of Durham in the UK, focusing on the prehistory of Europe and southwest Asia. From 1988 to 1989, he taught at the Memorial University of Newfoundland, Canada. After reading archeology at Cambridge University, Dr Rowley-Conwy completed his PhD in 1980 on the prehistory of Denmark, and has specialized in the analysis of prehistoric animal bones and plant remains. Since 1980 he has carried out post-doctoral work in Syria and has conducted research in Portugal, Italy, Egypt, and Sweden.

FRANK H. TALBOT
Dr. Talbot is Director of the National Museum of Natural History/National Museum of Man, Smithsonian Institution, Washington DC. A marine biologist who has specialized in coral reef fish ecology, he has studied reef fish communities off the east coast of Africa and off Australia's Great Barrier Reef. He has lived and studied on a Caribbean coral reef in the Tecktite II underwater habitat program, and sailed on a small yacht with his wife and youngest child across the Indian Ocean from Australia to Africa, returning via the Southern Ocean. Dr. Talbot has been Professor of Biology and Environmental Studies at Macquarie University, Sydney, Director of the Australian Museum in Sydney, and Executive Director of the California Academy of Sciences.

PAUL MICHAEL TAYLOR
Dr. Taylor is a research anthropologist and Director of the Asian Cultural History Program at the National Museum of Natural History, Smithsonian Institution, Washington DC. He received his Masters of Philosophy and PhD (both in anthropology) from Yale University. His main areas of study are folk biology, art and material culture, and the ethnography and languages of Southeast Asia. Dr. Taylor is a board member of the Association for Asian Studies (USA). He has written three books, some thirty publications, and has produced nine museum exhibitions.

DIANA WALKER
Dr. Walker is Senior Lecturer in Marine Botany at the University of Western Australia, where she has carried out research on seagrasses and macroalgae from the northwest of Western Australia to the south coast. She studied marine biology at the University of Liverpool's Marine Biological Station on the Isle of Man, and carried out her PhD research on coral reef algae in the Red Sea (Jordan) while at the University of York in England. Her main interests are the factors influencing the distribution of marine macrophytes covering ecophysiology to biogeography.

G.M. WELLINGTON
Dr. Wellington is Associate Professor of Biology, University of Houston, Texas, where he has studied the ecology and evolution of coral reef organisms. He is currently working on a project to evaluate the relationship between planktonic larval duration and gene flow in marine shorefishes of the tropical eastern Pacific. Between 1973 and 1975 he worked for the Charles Darwin Foundation on the Galapagos Islands and the Galapagos National Park Service (Ecuador) on a survey of marine environments and developed a comprehensive plan for their protection.

JOHN C. YALDWYN
Dr. Yaldwyn retired as Director of the National Museum of New Zealand in 1990, where he was based for 23 years. A specialist in marine biology, he worked at the University of Southern California's Allan Hancock Foundation in Los Angeles, and at the Australia Museum, Sydney. He has researched the biology and ecology of many islands, including New Zealand subantarctic islands, Great Barrier Reef and Coral Sea islands, and the raised coral atoll of Niue in the South Pacific. His publications include *Australian Seashores in Colour* and *Australian Crustaceans in Colour*, and he is now an honorary research associate of the National Museum of New Zealand.

ACKNOWLEDGMENTS

The publishers would like to thank the following people for their assistance in the production of this book: Michelle Boustani, Sue Burk.
The illustrations and maps in this book were produced by the following illustrators:
Anne Bowman 23, 50, 71, 86
Greg Campbell 96 (silhouette)
Jon Gittoes 76–7, 96–7
Mike Gorman 16, 26, 27, 36, 37, 38, 46, 60, 87, 100, 112–3, 130, 135
David Kirshner 32, 33, 75, 79, 92, 98, 106
Nicola Oram 48–9, 61, 72
Oliver Rennert 20–1, 43, 52–3, 63
Patrick Watson 89

Many of the illustrations prepared for this publication were based on original reference provided by the contributors. Other sources of illustrations are listed below. **Page 23** Reference for *Star sands* provided by Dr. Collins and Zoli Florian, James Cook University of North Queensland, Australia. **Page 36** *The making of oceanic volcanoes* is adapted from *Volcano* 1982, Time-Life Books, Amsterdam, pages 62–63. **Page 43** *Inside Kilauea* is adapted from "Dymanics of Kilauea Volcano," by John J. Dvorak, Carl Johnson, and Robert I. Tilling, © August 1992, international edition by Scientific American, Inc. All rights reserved. **Page 46** *Darwin's model* is adapted from Davis W.M. 1928, *The Coral Reef Problem*, American Geographical Society. **Page 86** *Rail skeleton* is adapted from Olson, Storrs, 1973, *Evolution of the South Atlantic Islands (Aves: Rallidae)*, Smithsonian Institution, p. 34

INDEX

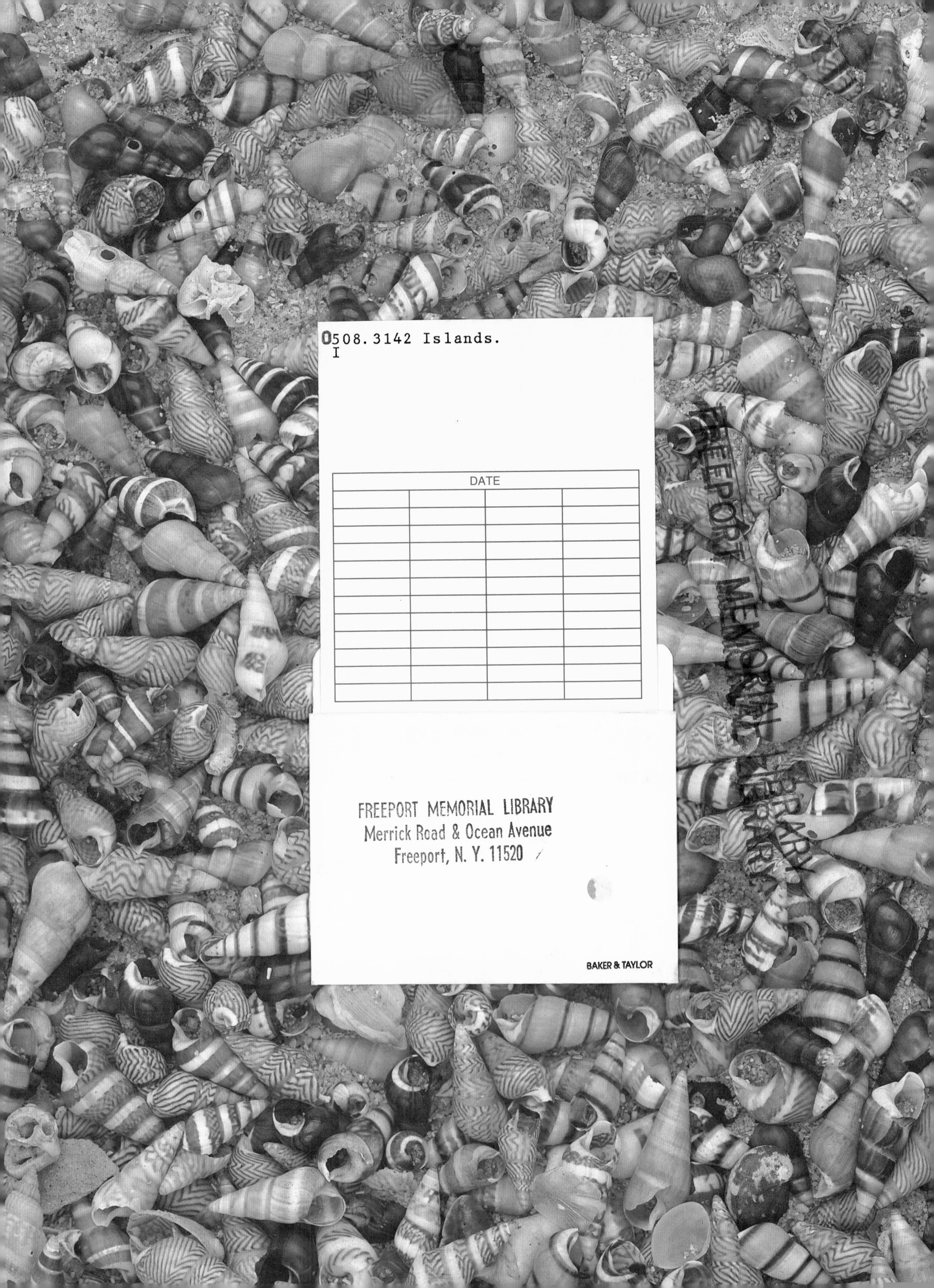

0508.3142 Islands.
I
DATE
FREEPORT MEMORIAL LIBRARY
Merrick Road & Ocean Avenue
Freeport, N. Y. 11520
BAKER & TAYLOR
FREEPORT MEMORIAL LIBRARY